ADVANCES IN QUANTITATIVE STRUCTURE–PROPERTY RELATIONSHIPS

Volume 3 • 2002

ADVANCES IN QUANTITATIVE STRUCTURE–PROPERTY RELATIONSHIPS

Editors: MARVIN CHARTON
BARBARA I. CHARTON

Volumes 1 and 2 were published by:

JAI PRESS INC.
100 Prospect Street
Stamford, Connecticut 06904-0811
USA

ADVANCES IN QUANTITATIVE STRUCTURE–PROPERTY RELATIONSHIPS

Editors: MARVIN CHARTON
Department of Chemistry
Pratt Institute
Brooklyn, New York

BARBARA I. CHARTON
Coles Science Center
New York University Library

VOLUME 3 • 2002

2002

JAI

An Imprint of Elsevier Science
Amsterdam – Boston – London – New York – Oxford – Paris
San Diego – San Francisco – Singapore – Sydney – Tokyo

ELSEVIER SCIENCE B.V.
Sara Burgerhartstraat 25
P.O. Box 211, 1000 AE Amsterdam, The Netherlands

ISBN: 0-444-51112-1
First edition 2002

British Library Cataloguing in Publication Data
A catalogue record from the British Library has been applied for.

Library of Congress Cataloging in Publication Data
A catalog record from the Library of Congress has been applied for.

∞ The paper used in this publication meets the requirements of ANSI/NISO Z39.48–1992 (Permanence of Paper)

Printed in The Netherlands.

CONTENTS

LIST OF CONTRIBUTORS

Marvin Charton	Chemistry Department School of Liberal Arts and Sciences Pratt Institute Brooklyn NY 11205 USA
Roger Gallo	ENSPICAM, UMR 6516 CNRS Fac Sciences St Jerome Av Normandie-Niemen 13397 Marseille Cedex 20 France
Stawomir J. Grabowski	Department of Crystallography and Crystallochemistry University of Lódz ul. Pomorska 149/153 90-236 Lódz Poland
S. P. Gupta	Department of Chemistry Birla Institute of Technology and Science Pilani-333031 India
Guy Jacob	SNPE Propulsion Centre de Recherches du Bouchet 9 Rue Lavoisier 91710 Vert Le Petit France
Tadeusz M. Krygowski	Department of Chemistry University of Warsaw L. Pasteura 1 02-093 Warsaw Poland

Guy Piacenza
ORSYMONDE Synthèse Organique
Ingénieur Chimiste R&D/Chemist Engineer
Z. I. de Mitry-Compans
Rue Gay Lussac – BP 230
77 292 Mitry Mory Cedex
France

Paul G. Seybold
Departments of Chemistry and Biochemistry
Wright State University
Dayton, OH 45435
USA

PREFACE

Over the last forty years quantitative structure property relationships (QSPR) have developed into a major method of chemical research. In the course of this development, the field has suffered from fragmentation. Thus, those working in physical organic chemistry form one constituency, those in inorganic chemistry another, spectroscopists constitute a third, medicinal and agricultural chemists a fourth, environmental chemists are yet another, and so on. There is nevertheless a common thread which runs through these widely diverse areas. The basic principles, parameterization, and methodology are the same or similar throughout.

The object of this series is to provide interesting and timely reviews covering all aspects of the field. It is our hope that this will encourage the transfer of new methods, techniques, and parameterizations from the field in which they were developed to other areas that can make good use of them. There is too often at present a tendency for workers in one discipline to be unaware of developments in other disciplines which might conceivably be of great utility. In view of the fact that QSPR have spread throughout the entire domain of modern chemistry, this is hardly surprising.

We hope that this series will provide the cross fertilization which we believe to be so sorely needed.

Marvin Charton

ELECTRONIC EFFECTS OF SUBSTITUENTS ON OXIDATION REACTIONS

S. P. Gupta

Advances in Quantitative Structure Property Relationships
Volume 3, pages 1–26.

I. INTRODUCTION

Today it is well known that the variations in the chemical or biological reactivity of compounds depend upon the variations in their structures, and it has long been held that similar changes in structure produce similar changes in reactivity. In a series of substituted analogues, therefore, the reactivity may depend upon the nature of the substituents. Substituents can affect the reactivity broadly by their electronic or steric properties. In oxidation and reduction like reactions, which involve electron or proton transfer, the electronic properties of the substituents seem to play a dominant role.

Around 1935, L. P. Hammett[1] made a notable contribution, parametizing the electronic effect. He proposed that the effect of substituents on the ionization of benzoic acids could be used as a model system to estimate the electronic effect of substituents on similar reaction systems. For example, a parallelism should exist between the ionization constant (K_a) of substituted benzoic acids and that of substituted phenyl acetic acids. The most surprising was, however, that even the rate constants (k) of substituted compounds undergoing any chemical reaction was found to be linearly correlated with K_a of the corresponding series of benzoic acids. These observations became the basis to develop several electronic parameters. To give the idea of various electronic parameters, we start here with the Hammett equation.

A. Hammett Equation

To directly measure the electronic effect of substituents, Hammett defined an electronic parameter σ as:

$$\sigma = \log K_X - \log K_H \qquad (1)$$

where K_X is the ionization constant of a *meta*- or *para*-substituted benzoic acid in water at 25 °C and K_H is the ionization constant of parent benzoic acid under the same experimental conditions. In terms of the rate constants, Eq. 1 could be written as:

$$\log k_X = \rho\sigma + \log k_H \qquad (2)$$

where the slope ρ (referred to as the reaction constant) is the measure of the sensitivity of the reaction to the electronic effect of the substituents. The positive values of σ represent electron withdrawal by the substituents from the aromatic ring and the negative values represent electron release by the substituents to the ring, both relative to H.

The Hammett equation (Eq. 2) has been well established to hold good for almost all types of organic reactions and more so when the reaction center is insulated from resonance interaction with the aromatic ring (e.g. X—$C_6H_4CH_2Q$, where Q is the reaction center). In the use of the Hammett equation, the conformation is neglected. The conformation of substituents is assumed to be taken care of by the model system (i.e. benzoic acids), but that of reaction centers is completely ignored. The conformation of a huge variety of reaction centers can, however, be accounted for by considering the resonance effects for which σ^- and σ^+ have been defined.

B. Electronic Effects Through Resonance

The ionization or other reactions of compounds where the reaction centers were not insulated from the aromatic ring, e.g. phenols, were not found so well correlated with σ, as the direct conjugation of substituent with the reaction center—the resonance phenomenon—as shown for phenol, as an example, affected the reactivity of the compounds. For such situations, σ^- and σ^+ were defined for the substituents taking phenols or anilines as a

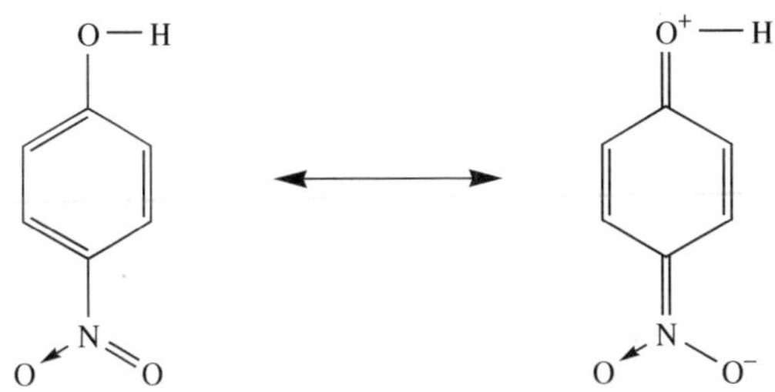

model system for σ^- and cumyl chlorides as a model system for σ^+. σ^- is meant for the substituents that can withdraw the electrons from, and σ^+ for the substituents that can donate the electrons to, the reaction center through resonance. However, both σ^- and σ^+ have been derived for only *para* substituents. For meta substituents, both are essentially same as σ_m, because the resonance effect from the meta position is not so significant.

C. Resonance and Field (or Inductive) Effects

The Hammett σ constant appears to be a combination of two electronic effects: resonance effect σ_R (or simply expressed as *R*) and inductive effect σ_I, also called as field effect and expressed as σ_F or simply as *F*. According to Swain and Lupton,[2] therefore, σ can be expressed as:

$$\sigma = aR + bF \qquad (3)$$

But to evaluate *F*, these authors surmised that *F* was a component in both σ_m and σ_p and therefore proposed to evaluate it by the regression analysis:

$$F = a\sigma_m + b\sigma_p + c \qquad (4)$$

Extending this approach to calculate *R* and *F* values for more than 500 substituents, Hansch et al.[3] formulated Eq. 5 for a set of 38 compounds for which *F* values were estimated from the ionization constants of 4-substituted bicyclo[2.2.2]octanecarboxylic acids where the resonance contribution of substituents was supposed to be zero. In this equation, *n* is the number of data points, *r* is the correlation coefficient, and *s* is the standard deviation.[4]

$$F = 1.297\,\sigma_m - 0.385\,\sigma_p + 0.033$$
$$n = 38 \qquad r = 0.968 \qquad s = 0.046 \qquad (5)$$

After the calculation of *F* values, *R* values can be calculated from the equation:

$$R = \sigma_p - F \qquad (6)$$

For a ready reference, a table is given (Table 1) that lists the values of *R* and *F* and other electronic parameters discussed for some important substituents.

All the electronic parameters have been usually evaluated for *para* and *meta* substituents. Ortho substituents pose special problems because their σ values include a steric contribution. Further, as compared to *meta*- and *para*-substituents, many *ortho*-substituents cause conformational changes, which can be favorable also and unfavorable also for the recactivity of the compounds.

Aliphatic σ constants (σ^*) are defined by Eq. 7, where k_X is the rate constant for the hydrolysis of the substituted acetates (X—CH_2COOR; *R*=Me or Et) and k_H is the rate constant for the hydrolysis of the parent acetate (X=H). The subscripts A and B refer to the acidic and basic conditions.

$$\sigma^* = (1/2.48)[\log(k_X/k_H)_B - \log(k_X/k_H)_A] \qquad (7)$$

Table 1. Electronic Parameters of Some Selected Aromatic Substituents[a]

No.	*Substituent*	σ_m	σ_p	σ_p^o	σ_p^+	σ_p^-	*R*	*F*
1	Br	0.39	0.23	0.30	0.15	0.25	−0.22	0.45
2	Cl	0.37	0.23	0.28	0.11	0.19	−0.19	0.42
3	F	0.34	0.06	0.21	−0.07	−0.03	−0.39	0.45
4	SF_3	0.70	0.80	—	—	—	0.17	0.63
5	SF_5	0.61	0.68	—	—	0.86	0.12	0.56
6	I	0.35	0.18	0.31	0.14	0.27	−0.24	0.42
7	NO_2	0.71	0.78	0.82	0.79	1.27	0.13	0.65
8	N_3	0.37	0.08	—	—	0.11	−0.40	0.48
9	H	0.00	0.00	0.00	0.00	0.00	0.00	0.00
10	OH	0.12	−0.37	−0.16	−0.92	−0.37	−0.70	0.33
11	NH_2	−0.16	−0.66	−0.36	−1.30	−0.15	−0.74	0.08
12	SO_2NH_2	0.53	0.60	—	—	0.94	0.11	0.49
13	CF_3	0.43	0.54	0.54	0.61	0.65	0.16	0.38
14	OCF_3	0.38	0.35	—	—	—	−0.04	0.39
15	SO_2CF_3	0.86	0.96	0.93	—	1.63	0.22	0.74
16	CN	0.56	0.66	0.68	0.66	1.00	0.15	0.51
17	CHO	0.35	0.42	0.54	0.73	1.03	0.09	0.33
18	$CONH_2$	0.28	0.36	—	—	0.61	0.10	0.26
19	CH_3	−0.07	−0.17	−0.12	−0.31	−0.17	−0.18	0.01
20	$NHCONH_2$	−0.03	−0.24	—	—	—	−0.33	0.09
21	OCH_3	0.12	−0.27	0.11	−0.78	−0.26	−0.56	0.29
22	$SOCH_3$	0.52	0.49	0.57	—	0.73	−0.03	0.52
23	SO_2CH_3	0.60	0.72	0.75	—	1.13	0.19	0.53
24	SCH_3	0.15	0.00	0.08	−0.60	0.06	−0.23	0.23
25	C≡CH	0.21	0.23	0.23	0.18	0.53	0.01	0.22
26	$CH{=}CH_2$	0.06	−0.04	0.00	−0.16	—	−0.17	0.13
27	$COCH_3$	0.38	0.50	0.50	—	0.84	0.17	0.33
28	$COOCH_3$	0.36	0.45	0.46	0.49	0.64	0.11	0.34
29	$NHCOCH_3$	0.21	0.00	−0.09	−0.60	−0.46	−0.31	0.31
30	$N(CH_3)_2$	−0.16	−0.83	−0.48	−1.70	−0.12	−0.98	0.15
31	Cyclopropyl	−0.07	−0.21	−0.10	−0.41	−0.09	−0.23	0.02
32	$CH(CH_3)_2$	−0.04	−0.15	−0.16	−0.28	−0.16	−0.19	0.04
33	$CH{=}C(CN)_2$	0.66	0.84	0.80	0.82	1.20	0.28	0.57
34	$C(CH_3)_3$	−0.10	−0.20	−0.17	−0.26	−0.13	−0.18	−0.02
35	C_6H_5	0.06	−0.01	0.04	−0.18	0.02	−0.13	0.12
36	$N{=}NC_6H_5$	0.32	0.39	0.36	−0.19	0.45	0.09	0.30
37	OC_6H_5	0.25	−0.03	0.08	−0.50	−0.10	−0.40	0.37
38	NHC_6H_5	−0.02	−0.56	−0.27	−1.40	−0.29	−0.78	0.22
39	COC_6H_5	0.34	0.43	0.50	0.51	0.83	0.12	0.31
40	$CH_2C_6H_5$	−0.08	−0.09	—	−0.28	−0.09	−0.05	−0.04
41	$C{\equiv}CC_6H_5$	0.14	0.16	—	−0.03	0.30	0.01	0.15
42	$CH{=}CHC_6H_5$	0.03	−0.07	—	−1.00	0.13	−0.17	0.10

Note: [a] Taken from Hansch, C.; Leo, A. *Exploring QSAR: Fundamentals and Applications in Chemistry and Biology*, American Chemical Society: Washington, D.C, 1995.

II. OXIDATION REACTIONS AND ELECTRONIC EFFECTS

The chemical mechanisms of oxidation of many compounds have been well studied and in many cases the substituents have been found to have a profound effect on the rate of the reactions.

The oxidation has been defined in many ways. The most prominent definitions, however, refer to addition of oxygen, removal of hydrogen, or transfer of electrons, of which the last one is the most favored one at the present time. Thus, since the oxidation is defined in terms of electron transfer, the electronic nature of the substituents does affect the rate of the oxidation reactions. Although an oxidation necessarily involves a reduction, attentions have been mostly paid to those reactions in which an organic substrate reacts with an oxidizing agent, often an inorganic molecule or metal ion. The use of metal ions as oxidants has received great attention.[5,6] Metal ions are involved in either one-electron transfer or two-electron transfer from the substrate to the metal ions. One of the tools with which to differentiate one-electron transfer from two-electron transfer is the magnitude of the reaction constant ρ. In one-electron transfer reactions, the ρ value is very low,[7–12] as in them the electronic charge separation between the initial and transition states is very small. In two-electron transfer reactions, the ρ value is usually quite high.[13–18]

The one-electron or two-electron transfer reactions can also be brought by the use of inorganic molecules/ions, such as selenium dioxide, dimethyl sulfoxide (DMS), oxides and acids of nitrogen, quinones, carbonium ions, peroxides, etc., and can be characterized by the ρ values. The purpose of this article is to characterize the ρ values for the varying types of oxidation reactions and to describe the comparative substituent electronic effects on them. Since the purpose of this article is not to compile all oxidation reactions studied, as they cover a vast amount of literature, a few selected examples have been taken. Many thousands of papers cover such an enormous range of applications of Hammett–Taft's equations that it would be virtually impossible to tabulate all of them, but the examples of particular interest have been discussed from time to time.[13,19–23] However, in the recent past there have been no such attempts.

A. Aromatic Amines

The oxidations of aromatic amines (anilines) have been studied using variety of oxidants. Sharadamani and Jaganadham[24] used *N*-bromoacetamide to study the kinetics and mechanism of a small set of compounds (aniline, *p*-toluidine, *p*-chloroaniline, and *m*-nitroaniline). The reaction took place according to Eq. 8 and the major product formed was found to be 4,4′-dinitro azobenzene.

$$2C_6H_5NH_2 + 2CH_3CONHBr \rightarrow C_6H_5{-}N{=}N{-}C_6H_5 + 2CH_3CONH_2 + 2HBr \quad (8)$$

The rate of the reaction was found to depend on the electronic nature of the substituents in the aromatic ring. The electron-releasing substituents were observed to increase the rate of the reaction whereas the electron-withdrawing ones to decrease it. A highly significant linear correlation (correlation co-efficient $r = 0.998$) was found to exist between the rate constant (taken as log k) at 303 K and the σ values of the substituents with a slope (ρ value) of -0.785. This value of ρ is slightly less than those reported for the oxidation of anilines by some other workers.[25–27] However, similar ρ values have been reported for the oxidation of benzaldehydes and bromination of phenols by *N*-bromoacetamide.[28,29]

The rate of oxidation of *N*,*N*-dimethyl anilines by peroxodisulphate was also found to be increased by electron-releasing substituents and to be decreased by electron-withdrawing substituents.[30] For a small series of *meta*- and *para*-substituted *N*,*N*-dimethyl anilines, Srinivasan et al.[30] had found an exellent correlation existing between the logarithms of second order rate constants and σ^+ ($\rho^+ = -0.765$, $r = 0.998$). Though the slope in this case appears to be almost identical to that in the preceding case, the correlaion with σ^+ indicates that the substituents produce their effect through direct resonance.

The reaction between aromatic amines and peroxodisulphate (the Boyland–Sims oxidation) gives ortho-aminoaryl sulphates (**1**).[31–33] In a study on the oxidation of ortho-substituted *N*,*N*-dimethyl anilines by peroxodisulphate, Srinivasan et al.[34] found that ortho substituents also increase the rate of reaction if they are electron-releasing, but their effect is

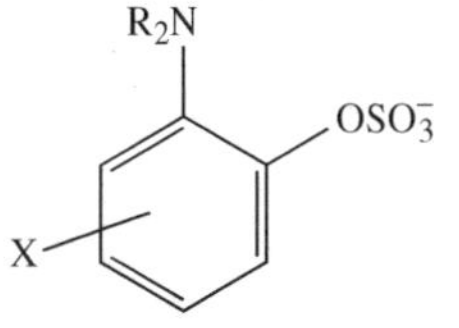

1 (R = Me or H)

through both inductive (field) and resonance effects (Eq. 9).[34] However, unlike the *meta*- and *para*-substituents, the *ortho*-substituents were found to produce also the rate-retarding steric effect[34] as accounted for by a volume parameter ν in Eq. 9.

$$\begin{aligned} \text{Log k} &= -1.91\,\sigma_I - 0.682\,\sigma_R - 2.13\,\nu - 1.4565 \\ &\quad (\pm 0.291) \quad (\pm 0.309) \quad (\pm 0.377) \\ n &= 7 \quad r = 0.985 \quad F = 33.4 \end{aligned} \tag{9}$$

This rate-retarding effect of *ortho*-substituents led Srinivasan et al. to confirm their and Behrmans' earlier assumption that oxidant attacks *ipso*-carbon (C1) and not the *ortho*-carbon or amine nitrogen.[30,35]

In Eq. 9, the figures within parenthesis below each variable refer to confidence levels of student's T-test—$>$ 99% for σ_I, $<$90% for σ_R, and $>$ 98% for ν. F is the F-ratio between the variances of calculated and observed activities.

The electron-releasing effect of substituents on oxidation of anilines has been shown with other oxidants also. In a study on hexacyanoferrate (III) – catalyzed perborate oxidation of anilines by Karunakaran and Palaniswamy,[36] the rates of the reactions for *para*-, *meta*- and *ortho*-substituted analogues were found to be related with electronic parameters as:

para-substituted analogues:

$$\begin{aligned} \log k^* &= -\,0.78\,\sigma_I - 0.91\,\sigma_R^- + \text{constt} \\ n &= 10 \quad r^2 = 0.898 \end{aligned} \tag{10}$$

meta-substituted analogues:

$$\begin{aligned} \log k^* &= -0.79\,\text{F} - 0.47\,\text{R} + \text{constt} \\ n &= 8 \quad r^2 = 0.970 \end{aligned} \tag{11}$$

ortho-substituted analogues:

$$\begin{aligned} \log k^* &= -\,1.69\,\text{F} - 0.52\,\text{R} - 0.50\,\nu + \text{constt} \\ n &= 7 \quad r^2 = 0.973 \end{aligned} \tag{12}$$

All the above three equations exhibit the electron-releasing effect of the substituents, but through inductive (or field) and resonance phenomena. The *ortho*-substituents are again found to produce rate-retarding steric effect. The rate constant k^* was derived from $k^* = k'/\,[\text{aniline}]_o$, where k' is the pseudo first-order rate constant. It was evaluated at 4 different temperatures, *viz*., 40, 45, 50, and 55 °C. We have cited here the correlations of only 40 °C. Correlations of other temperatures were almost parallel and of same significance. The constants (the intercepts) were not reported for any equation.

The rate of oxidation of diphenylamines (**2**) by iodate, leading to the formation of carbazole (**3**), was shown[37] to be correlated with the simple Hammet constant (σ) with a value of $r = 0.98$. While the r-value remaining almost same, the ρ-values obtained at three different temperatures, 293, 298 and 303 K were found to be -6.25, -6.07 and -5.96, respectively, exhibiting high effect of electron-releasing substituents on the rate of the reaction. The substituents studied were H, 3-CH_3, 3-Cl, 4-SO_3H, and 3-CH_3O.

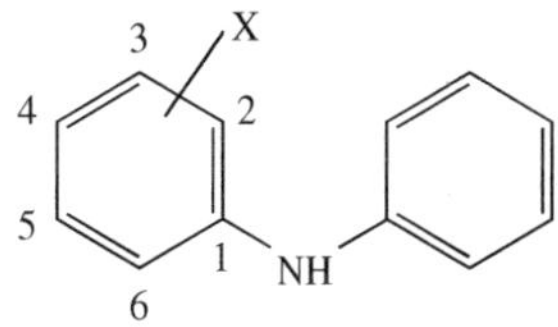

2

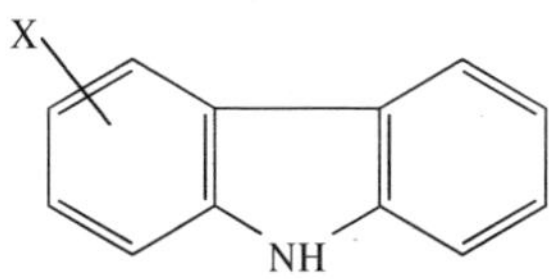

3

Assuming that attack of the oxidant on the amine was at position 2 (ortho to the amino group), Brahmaiah et al.[37] had used the σ-values of substituents relative to this position. The existence of a high correlation between $\log k_{obs}$ and σ so used led these authors to confirm their assumption. Similar studies were made for the oxidations of phenols also by iodate to confirm the attack of oxidant at the position ortho to phenolic hydroxyl group.[38]

B. Phenols

The oxidation of phenols has been studied using several kinds of oxidants. The oxidation with iodate has drawn relatively more attention and several authors specifically studied the electronic effects of substituents on the rate of the oxidation.[39–41] Manikyamba et al.[39] studied the oxidation of phenols by iodate in aqueous methanol. With respect to both substrate (phenols) and oxidant (iodate), the reaction was found to be of first order and the rate constant was shown to be significantly correlated with σ with a value of $\rho = -5.2$, suggesting that electron-releasing substituents would have a significant positive effect on the rate of the oxidation. Similar observations were made by Radhakrishna Murthy and Janardhan[40] in the iodination of phenols, but in their study ρ was found to have a very low value equal to -1.6.

The oxidation of phenols by periodate was found by Radhakrishna Murthy et al. to involve the same mechanism as iodate oxidation.[41] Almost identical vaues of ρ^+, -4.0 and -4.5, were found in both the cases[41] when $\log k$ was correlated with σ^+.

The electron-releasing effect of substituents on the rate of oxidation was also observed when acid bromate was used as oxidant.[42] The log k was found to be linearly related with σ with a value of $\rho = -1.74$.

In a study of Ag^{+}– catalyzed oxidation of phenols by peroxydisulphate, Sharma et al.[43] showed that a good correlation between log k and σ could be obtained only when the σ-values were taken relative to the position of phenolic OH. Though the values of $r = 0.79$ and $\rho = -0.32$ were not very high, these authors suggested that electron-releasing substituents would increase the rate of oxidation and the electron-withdrawing ones would decrease it and that the reaction center would be the oxygen of phenolic group. The attack of oxidant at phenolic oxygen in phenols was also suggested in iodate and periodate oxidations in acetic acid medium[41] as well as in acid bromate oxidation.[42] However, in aqueous methanol medium, the attack of IO_3^- was suggested to be at the carbon atom, *ortho* to phenolic OH.[39]

Reddy and Jagannadham[44] studied the kinetics and mechanism of oxidation of phenols by chromic acid in acetic acid-water mixture and noted the favorable effect on the rate of the reaction of electron-releasing substituents. The logarithm of the pseudo first- order rate constant k′ was found to be correlated with σ^+ with a value of $r = 0.9812$ and $\rho^+ = -3.36$ at 297 K.

For the ortho substituents, a rate retarding steric effect was also observed by Reddy and Jagannadham[44] as expressed by the equation:

$$\mathrm{Log}(k''/k''_o) = -2.81\sigma^* + 0.477\,E_s + \mathrm{constt} \tag{13}$$

In this equation, E_s is Taft's steric parameter and k'' and k_o'' represent the biomolecular rate constants for the chromic acid oxidation of o-substituted phenols and o-cresol, respectively.

A mechanism involving a direct attack of $HCrO_3^+$ on oxygen atom of phenol in rate determining step was proposed by these authors and the end product was found to be o-benzoquinone, which has been usually the product of all oxidations of phenols.

C. Benzaldehydes

The iodates were used to oxidize the benzaldehydes also. Manikyamba et al.[45] studied the kinetics and mechanism of oxidation of benzaldehydes by acid iodate in aqueous methanol. The reaction takes place according to the equation:

$$3C_6H_5CHO + IO_3^- \rightarrow 3C_6H_5COOH + I^- \tag{14}$$

Regarding the electronic effect of the substituents on such oxidation reaction, Mankyamba et al. found that the plot of log k vs σ was not a straight line but a smooth curve (inverted parabola), suggesting that electron-releasing substituents will have an optimum effect on the rate of the reaction. Similar curves were obtained when σ^+ or σ^- were used in place of σ. Such type of curve was also obtained in various other reactions, e.g. reactions of benzyl chlorides with different nucleophiles in several solvents,[46] reactions of anilines with benzaldehyde in ethanol medium,[47] solvolysis of benzenesulphonyl chlorides in water.[48]

In periodate oxidation of benzaldehydes, catalyzed by OsO_4^- in alkaline medium, Panigrahi and Misra[49] observed that at a particular OH^- concentration all the substituted benzaldehydes with electron-releasing and electron-withdrawing substituents reacted faster than the unsubstituted one and that a plot of log k_{obs} vs σ showed a break indicating either a breakdown of $\rho - \sigma$ relationship or that different mechanisms were operative for substrates having electron-releasing and electron-withdrawing substituents. However, as OH^- concentration was increased, the break in log k_{obs} vs σ plot was found to vanish.

In the oxidation of benzaldehydes by chromic acid, Musala Reddy et al.[50] observed that the plot of log k vs σ for *meta*-substituents was different from that for *para*-substituents. For the former they found $r = 0.970$ and $\rho_m = 0.851$ and for the latter $r = 0.989$ and $\rho_p = 1.23$. This difference in the correlation and a higher value of ρ_p led Musala Reddy et al. to suggest that the *para*-substituents might have the resonance effects. However, the positive values of ρ_m and ρ_p suggested that the electron-withdrawing and not the electron-releasing substituents would favor the reaction. This suggestion was made by Lucchi also in a number of studies,[51–53] and using Lucchi's data, Graham and Westheimer[54] had obtained $\rho = 1.06$. On the basis of the positive value of ρ the rate-controlling step was postulated as the loss of the aldehydic hydrogen atom as a proton. In a more careful examination of substituents effects on chromic acid oxidation of benzaldehydes, Wibergand and Mill[55] were able to correlate log k with σ with $r = 0.96$ and $\rho = 1.14$ ($n = 8$).

That the rate of the oxidation of benzaldehydes is increased by electron-withdrawing substituents and not by electron-releasing ones was also shown by Reddy et al.,[56] who studied the oxidation of some *meta*- and *para*-substituted benzaldehydes with Ni(III) in acetic acid–water mixture and found log k to be correlated with σ with $r = 0.978$ and $\rho = 1.92$ ($n = 7$). On the basis of this, Reddy et al. also assumed that the rate controlling step involved the loss of H, which could be facilitated by increase in the positive charge on the carbonyl carbon by electron-withdrawing substituents. Thus, a heterolytic fission of C—H bond of aldehyde group was proposed in the oxidation of benzaldehydes.

However, in V(V) oxidation of benzaldehydes, Murty et al.[57] had correlated the rate of the reaction with σ with a value of $\rho = -0.68$ and $r = 0.976$ ($n = 10$), suggesting that the rate of the reaction can be increased by electron-releasing substituents and not by electron-withdrawing ones. From this, these authors had assumed a free radical mechanism in which the rate-determining step produces a free radical ($Ar\dot{C}O$), which is then involved in the next step in the formation of a transition state. This transition state is very weak. The electron-releasing substituents increase the charge on the carbonyl carbon and stabilize the transition state, allowing the reaction to move in the forward direction.

A study on *N*-bromoacetamide oxidation of benzaldehydes in acetic acid–water mixture[58] had also revealed that electron-releasing substituents will increase the rate of oxidation and electron-withdrawing ones will decrease it, as the rate constant was found to be correlated with σ^+ with $r = 0.97$ and $\rho^+ = -1.675$ ($n = 5$).

However, the study of the kinetics of the oxidation of benzaldehydes by quinolinium dichromate corroborated the favour of the rate of reaction by electron-withdrawing substituents.[59] In their study, Aruna et al.[59] correlated $\log k_{obs}$ with σ with $\rho = 1.20$ and $r = 0.97$ ($n = 8$).

D. Aromatic Anils

Aromatic anils (**4**) have two phenyl rings X and Y. The X originates from benzaldehyde and Y from aniline. The anils are known to get cleaved oxidatively by lead tetraacetate,[60,61] phenyliodosoacetate,[62] and bromate,[63]

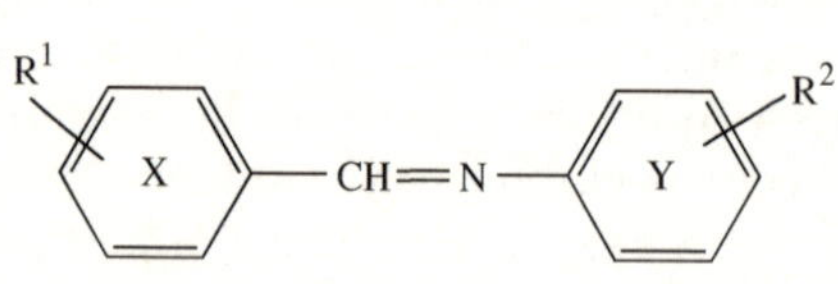

4

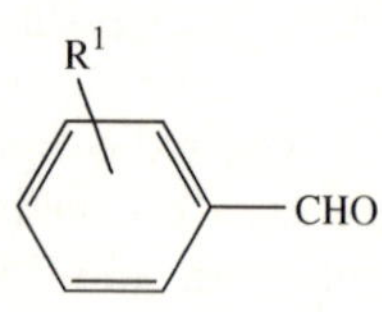

5

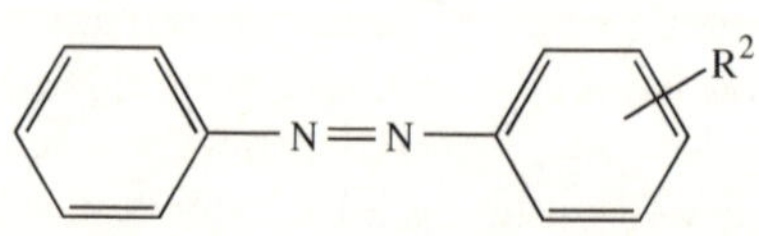

6

and the main products that result have been reported to be benzaldehydes (**5**) and azobenzenes (**6**).[60–63]

In a preliminary communication on phenyliodosoacetate oxidation of anils, Narasimhabharathi et al.[62] showed the rate of the reaction to be related with σ for the variation of the R^1-substituents on benzaldehyde ring (X) with $\rho = -0.68$ and $r = 0.97$ and for the variation of R^2-substituents on aniline ring (Y) with $\rho = 3.25$ and $r = 0.98$. Similarly, in a recent communication,[64] Karunakaran et al. found ρ for the variation of R^1-substituents ($R^2 = H$) equal to -0.64 and for the variation of R^2-substituents ($R^1 = H$) equal to 1.87. The value of ρ for the variation of the substituents in one ring, in fact, depends on the substituent present in the other ring. Such type of variation in the values of ρ has been analyzed in terms of interactive free-energy relationship (IFER) for multiple substituent effect as discussed by Ruasse and Argile.[65,66] According to these authors, the relation as given by Eq. 15 can be applicable for the reaction involving anil type of substrates, where $\rho_X(Y)$ means the value of ρ for the variation of the substituents in ring X with a particular substituent in ring Y and $\rho_Y(X)$ means the value of ρ for the variation of substituents in ring Y with a particular substituent in ring X. X_0 and Y_0 stand for H in the rings and σ_X and σ_Y are Hammett's constants for substituent in ring X and Y.

$$\frac{\rho_X(Y) - \rho_X(Y_0)}{\sigma_Y} = \frac{\rho_Y(X) - \rho_Y(X_0)}{\sigma_X} = q \qquad (15)$$

The q is supposed to be a constant and is referred to as the cross interaction constant. Equation 15 can be written as:

$$\rho_Y(X) - \rho_Y(X_0) = q_X \sigma_X \qquad (16)$$

and

$$\rho_X(Y) - \rho_X(Y_0) = q_Y \sigma_Y \qquad (17)$$

where q_X and q_Y are the cross interaction constants with respect to X and Y. Karunakaran et al.[64] found these equations quite applicable in their study by plotting $\rho_Y(X)$ vs σ_X and $\rho_X(Y)$ vs σ_Y and obtaining $q_X = 0.60$ and $q_Y = 0.55$. In both the cases, significant linear relationships were obtained with r equal to 0.992 and 0.994, respectively. The q_X and q_Y values were almost equal, validating Eq. 15 and suggesting that the attack of phenyliodosoacetate on the anil was a single concerted step and that the substituted anils were oxidized in each ring by the same mechanism as proposed by Narasimha Char et al. for the oxidation of anils by acid bromate.[67] In that, both the substrate (anils) and the oxidant (BrO_3^-) were supposed to undergo the protonation.

In the oxidation of anils by acid bromate, Narasimha Char et al.[67] had obtained $q_X = 0.70$ with $r = 0.986$, which is very close to $q_X = 0.60$ obtained by Karunakaran et al. for phenyliodosoacetate oxidation. However notwithstanding a negative value of ρ_R^+ with $R^2 = H$ in phenyliodosoacetate oxidation, a positive value of that equal to 0.76 was obtained by Narasimha Char et al. for acid bromate oxidation.

However, in a recent study on the kinetics and mechanism of oxidation of aromatic anils by *N*-chlorobenzamide, Mohamed Kasim and Basheer Ahamed[68] found $\rho = -0.40$ for the series with various substituents in ring X and hydrogen in ring Y, and $\rho = -2.65$ for the series with various substituents in ring Y and hydrogen in ring X. They also studied the correlations between $\rho_X(Y)$ and σ_Y and between $\rho_Y(X)$ and σ_X and found the linear relationships existing in both the cases with $q_Y = -1.46$ and $q_X = -1.51$, which essentially confirm the applicability of Eqs. 15–17. The value of r for the former was 0.99 and for the latter 0.98.

Since the values of q_X and q_Y were almost same, Mohamed Kasim and Basheer Ahamed also proposed that the attack of HOCl, the actual oxidizing agent formed by the presence of *N*-chlorobenzamide in aqueous methanol medium, on the double bond was a single concerted step. These similarities in the values of q_X and q_Y had led Padmaja et al.[69] also to assume the attack of OsO_4 on the double bond to be a single concerted step in Os(VIII)-catalyzed oxidation of phenyl stryl ketones (**7**) by periodate.

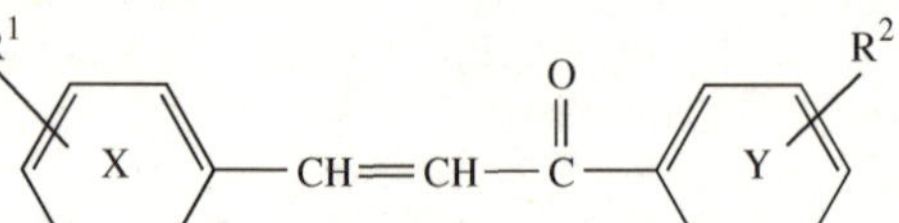

7

In their study on the oxidation of **7**, Padmaja et al. found $q_X = 0.41$ and $q_Y = 0.36$. In the Hammett plots, these authors had found $\rho_{R1} = -0.665$ with $R^2 = H$ and $\rho_{R2} = -0.947$ with $R^1 = H$. These negative values of ρ for the substituents at either ring in phenyl stryl ketones, as well as in anils studied by Mohamed Kasim and Basheer Ahamed, suggested that electron-releasing substituents at either ring would enhance the rate of oxidation and the electron-withdrawing ones will retard it. This suggestion, however, did not fully corroborate with the findings of Narasimhabharathi et al.[62] or of Karunakaran et al.[64] in phenyliodosoacetate oxidation of anils, where the ρ values for the substituents in ring Y were found to be positive suggesting that electron-releasing substituents at this ring will decrease the rate of oxidation. The suggestion also did not corroborate with the findings of Narsimha Char et al.[67] in acid bromate oxidation of anils, where ρ was found to be positive even for the substituents at ring X.

The q appears to be a sensitive parameter for predicting the site of attack. In the case of Ce(IV) oxidation of phenyl styryl ketones,[70] the values of q_X and q_Y were found to be −1.60 and −0.86, which were quite different from each other, and hence the site of attack was assumed to be the carbonyl group.

E. Sulfur-Containing Compounds

1. *Alkyl Aryl Sulfides*

In alkyl aryl sulfides, the phenyl methyl sulfides (**8**) and naphthyl methyl sulfides (**9**) have been relatively well studied for their oxidation. A variety of oxidants have been used to study the kinetics and mechanism of oxidation of these sulfides, and it has been found that the rates of oxidation of these compounds are largely affected by the nature of oxidants.

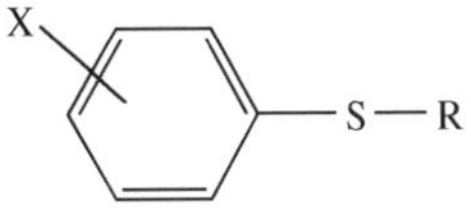

8

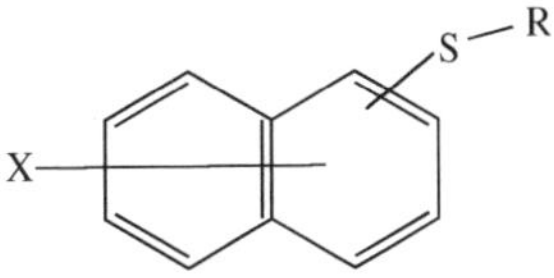

9

In a study on the oxidation of a series of *para-* and *meta-*substituted phenyl methyl sulfides by peroxydisulfate,[71] the rate constant k_2 (second-order) was found to be linearly correlated with σ^+/σ^- with $\rho^+ = -0.56$ and $r = 0.979$ ($n = 13$), suggesting that the rate of the reaction will be accelerated by the electron-releasing substituents and retarded by electron-withdrawing substituents and that the effect would be produced through direct resonance. The resulting product of oxidation of alkyl aryl sulfides are corresponding sulfoxides.

In an acid bromate oxidation of some *para-*substituted phenyl methyl sulfides, Rajasekaran et al.[72] had correlated the overall rate constant with σ_p and σ_p^- as:

$$\log k = -1.0134(\pm 0.03) - 1.41(\pm 0.61)[\sigma_p + 2.11(\pm 0.29)(\sigma_p^- - \sigma_p)] \quad n = 6 \quad r = 0.99 \qquad (18)$$

This equation also suggests that the electron-releasing substituents will enhance the rate of the reaction. Equation 18 was derived excluding all electron-withdrawing substituents.

Rajasekaran et al.[73] had also studied the oxidation of some *para-*substituted phenyl methyl sulfides by pyridinium chlorochromate in dipolar protic and aprotic solvents. In a protic solvent (aqueous acetic acid), they observed an overall second-order reaction and found the rate constant to be related with σ with $\rho = -2.12$, $r = 0.985$, and $s = 0.104$ ($n = 8$). In an aprotic solvent (chlorobenzene-nitrobenzene mixture), they correlated the rate of decomposition of the complex formed in the first step, which was rate-determining step, with σ with $\rho = -1.23$, $r = 0.989$, and $s = 0.068$ ($n = 8$). Thus, Rajasekaran et al. also observed that electron-releasing substituents will enhance the rate of oxidation and electron-withdrawing ones will decrease it.

Similar observations regarding the electronic effects of substituents were made in the case of the oxidation of naphthyl alkyl sulfides (**9**). For a series of 4-substituted 1-naphthyl methyl sulfides (**10**), the oxidation was studied with peroxoanions, using potassium peroxodisulphate (PS) and potassium peroxodiphosphate (PP).[74] In that study log k_2 (k_2 : second-order rate constant) was found to be highly correlated with σ_p^+ for both the oxidants (PS: $\rho^+ = -0.66$, $r = 0.999$, $s = 0.01$, $n = 8$; PP: $\rho^+ = -0.65$, $r = 0.994$, $s = 0.03$, $n = 8$). The ρ^+ values for both the oxidants are nearly equal. This is not surprising as the two oxidants are isoelectronic and isostructural with almost equal oxidation potential.[74]

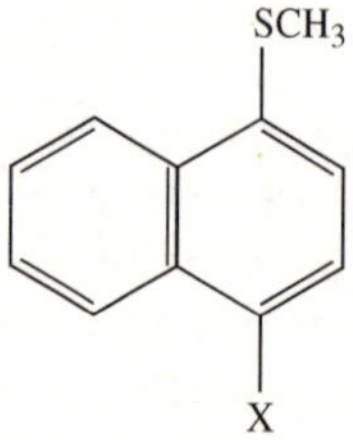

10

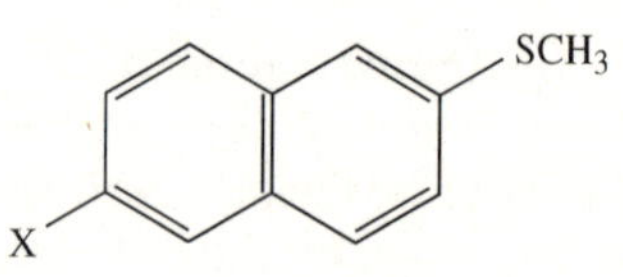

11

In a later study on the oxidation of 6-substituted 2-naphthyl methyl sulfides (**11**) by peroxoanions, Srinivasan et al.[75] found log k_2 for both oxidants to be better correlated with σ_p^+/σ_p^- than with σ_p or $\sigma_p{}^+$ [PS : $\rho(\sigma_p^+/\sigma_p^-) = -0.264$, $r = 0.987$, $s = 0.02$, $n = 6$; PP : $\rho(\sigma_p^+/\sigma_p^-) = -0.358$, $r = 0.996$, $s = 0.02$, $n = 6$]. The correlation of rate data with σ_p^+/σ_p^- in naphthyl methyl sulfides as well as in phenyl methyl sulfides indicate the ability of sulfur to donate or accept electrons through resonance effects. However, the negative sign of ρ values is in conformity with the well documented nucleophilic behaviour of sulfides in general.[71,76–78] The smallness of the ρ values suggests that a relatively "early" (reactant-like) transition state is involved in the rate-determining step.

For both **10** and **11** systems of naphthyl methyl sulfides, an oxidation study was made with Cr(VI).[79] In this study, while the rate data for 4,1-series (**10**) was found to be related with σ_p^+ ($\rho^+ = -1.72$, $r = 0.993$, $s = 0.09$, $n = 8$), for 6,2-series (**11**) it was found to be related with σ_p ($\rho = -1.18$,

$r = 0.991$, $s = 0.05$, $n = 8$). This difference in the behavior of substituents in the two series in Cr(VI) oxidation may be due to the variation in the location of the reaction site.

Recently the oxidation of a small series of diphenyl sulfides (**12**) by oxo(salen)manganese(V) [salen = *N*,*N*′-ethylene bis(salicylidineaminato)] complexes was studied by Chellamani and Alhaji.[80] Excellent correlation was shown to exist between log k_2 and σ_p with $\rho = -2.12$ and $r = 0.996$ ($n = 5$). A similar result was obtained by Chellamani et al.[81,82] for the oxidation of aryl methyl sulfides by oxo(salen)manganese(V) complexes and

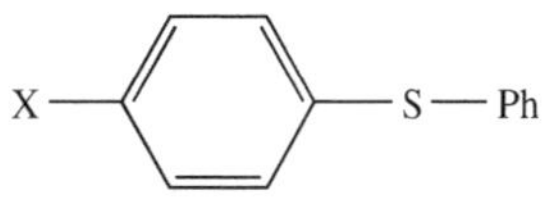

12

hence they proposed that diphenyl sulfides and aryl methyl sulfides are oxidized by this oxidant by a similar mechanism, where a single electron transfer can be assumed to take place, leading to the formation of a cation radical in the slow step.

2. *Alkyl Aryl Sulfoxides*

Sulfoxides are unique examples in oxidation reactions, as they may undergo both electrophilic and nucleophilic reactions.[83] Different oxidants have been used to study the oxidation of both alkyl aryl and diaryl sulfoxides. In a study on potassium bromate oxidation of phenyl methyl and diphenyl sulfoxides,[84] it was observed that for both the systems the reaction was of second order in aqueous acetic acid at constant pH and ionic strength (first order with respect to both substrate and oxidant) and in each case the rate constant (log k_2) was found to be significantly correlated with σ. For *meta*- and *para*-substituted phenyl methyl sulfoxides, the correlation was as shown by Eq. 19, and for some 4,4′-substituted diphenyl sulfoxides as shown by Eq. 20. Both Eqs.19 and 20 exhibit the favorable effect of electron-releasing substituents.

$$\log k_2 = -2.05(\pm 0.10)\sigma + \text{constt} \\ n = 12 \qquad r = 0.997 \qquad s = 0.05 \tag{19}$$

$$\log k_2 = -1.82(\pm 0.32)\sigma + \text{constt} \\ n = 6 \qquad r = 0.992 \qquad s = 0.12 \tag{20}$$

The oxidation of aryl methyl sulfoxides by oxo(salen)manganese(V) complexes was also shown to be favored by electron-releasing substituents.[85]

For a series of 6 *para*-substituted analogues, Chellamani et al.[85] had found log k_2 to be correlated with σ_p with $\rho = -2.44$, $r = 0.984$, $s = 0.079$. This value of ρ is almost similar to that in Eq. 19. A similar value of ρ was found for chloramin-T oxidation also ($\rho = -2.02$).[86] Several other oxidants had, however, given slightly lower values of ρ, e.g. dioxiranes,[87] gem-dialkylperoxonium ion,[88] peroxomonosulfate,[89] and oxo(phosphine)ruthenium(IV) complexes[90] had given ρ equal to −0.76, −0.83, −0.40, and −0.42, respectively. All these studies had, however, led the investigators to assume that the oxidation of sulfoxides, in most of the cases, involves an S_N2 mechanism with rate-determining electrophilic attack of the oxidant on the sulfur of the sulfoxide.[84–90]

For a series of diaryl sulfoxides ($n = 5$), an *N*-bromocaprolactam oxidation[91] had given $\rho = -1.34$, $r = 0.997$, $s = 0.07$, but in a Cr(VI) oxidation the result was slightly different in that the correlation was with Σ σ^+ and not with σ ($\rho^+ = -0.48$, $r = 0.989$, $s = 0.04$).[92]

A kinetic study[93] on Cr(VI) co-oxidation of several substituted phenyl methyl and diphenyl sulfoxides with oxalic acid has also revealed the dependence of rate constant on σ. For *meta*- and *para*-substituted phenyl methyl sulfoxides, the Hammett plot had $\rho = -0.927(\pm 0.08)$, $r = 0.994$, $n = 11$, and for 4,4′-substituted diphenyl sulfoxides, it had $\rho = -0.82(\pm 0.11)$, $r = 0.994$, $s = 0.06$, $n = 6$. In an oxidation of methyl phenyl sulfoxides by Cr(VI) in aqueous acetic acid studied by Baliah and Satyanarayanan,[94] the Hammett plot had given $\rho = -0.807$, $r = 0.969$, $s = 0.078$. This ρ value was comparable to the one obtained by Cerniani and Modena for the oxidation of aryl methyl sulfoxides with peroxybenzoic acid.[95]

3. *Phenylthioacetic Acids*

The studies on the kinetics and mechanism of phenylthio(or phenylmercapto)acetic acids ($X{-}C_6H_4SCH_2COOH$) or acetate ions($X{-}C_6H_4SCH_2COO^-$) are limited. However, in a study on potassium peroxy disulfate oxidation of phenylthioacetic acids, Srinivasan and Pitchumani[96] had found the rate of reaction (second-order) to be correlated with σ^+ ($\rho^+ = -0.618$, $r = 0.993$, $s = 0.041$, $n = 8$), showing that electron-releasing substituents will accelerate the rate of reaction and the electron-withdrawing substituents will decrease it and that the effect of substituents would be through direct resonance. The product formed was shown to be phenyl methyl sulfides, although Larsson[97] had claimed that the product should be phenylsulphinylacetic acids ($X{-}C_6H_5SOCH_2COOH$). However, in a later study on the oxidation of some *meta*- and *para*-substituted phenylthioacetic acids by bromamine T in alkaline medium,[98] Srinivasan et al. showed that the product could be only sulfinylacetic acids and not the methyl sulfides, but observed that the rate of reaction was related to σ and not to σ^+ (at 35 °C: $\rho = -2.46$, $r = 0.994$, $s = 0.01$, $n = 13$). This ρ value was exactly identical to the one found at the same temperature for the oxidation of phenylthioacetic acids by chloramine T($\rho = -2.46$).[99]

The negative value of ρ was obtained even in Mo(VI)-catalyzed perborate oxidation of phenylmercaptoacetic acids.[100] A small series of *para*- and *ortho*-substituted derivatives were studied[100] and at 30 °C the Hammett plot had given, for *para*-substituted derivatives, $\rho = -1.01$, $r = 0.98$, and $s = 0.09$ ($n = 6$) and, for *ortho*-substituted ones, $\rho^* = -0.83$, $r = 0.99$, and $s = 0.03$ ($n = 4$). The significant correlation of the rate constant with σ for *ortho*-substituted phenylmercaptoacetic acids indicates that probably *ortho*-substituents do not produce any steric effect.

Notwithstanding the results discussed so far, in a study on the oxidation of phenylmercaptoacetates by potassium hexa-cyanoferrate(III),[101] the value of ρ was found to be positive. The log k_2 for a series of 10 *para*- and *meta*-substituted analogues at 47 °C was found to be correlated with σ as :

$$\log k_2 = 1.31(\pm 0.09)\sigma - 2.55$$
$$n = 10 \qquad r = 0.997 \qquad s = 0.03 \tag{21}$$

This suggested that the rate of reaction will be accelerated by electron-withdrawing substituents and not by electron-releasing substituents. This led Srinivasan and Subramanian[101] to propose a different mechanism of oxidation, which involves an initial reversible proton abstraction followed by another reversible electron transfer. The removal of an acidic proton has been shown in a variety of hexacyanoferrate(III) oxidation.[102,103]

F. Amino Acids

Citing the example of amino acids here is just because of its biological importance. A study of the kinetics of oxidation of α-amino acids is important both from mechanistic point of view and its bearing on the mechanism of amino acid metabolism. Recently Pareek et al.[104] studied the oxidation of a few amino acids, namely glycine(Gly), α-alanine(Ala), valine(Val), leucine(Leu), phenylalanine(Phe), isoleucine(Ile), 2-aminobutanoic acid (ABA), norleucine(NLE), and norvaline(NVA), by *N*-chlorocarbamate in aqueous acetic acid and found the rate constant of the rate-determining step to be correlated with Taft's electronic and steric parameters of substituents at α-carbon as:

$$\log k = -2.55(\pm 0.43)\sigma^* - 2.81$$
$$n = 9 \qquad r = 0.912 \qquad s = 0.28 \tag{22}$$

$$\log k = -0.93(\pm 0.06)E_s - 3.02$$
$$n = 9 \qquad r = 0.984 \qquad s = 0.12 \tag{23}$$

These equations exhibit that not only the electron-releasing property of the substituents will favour the reaction but their steric property also would be advantageous. Both parameters were combined to give:

$$\log k = -0.89(\pm 0.07)\sigma^* - 0.65(\pm 0.02)E_s + \text{constt}$$
$$n = 9 \qquad r = 0.999 \qquad s = 0.02 \tag{24}$$

In an earlier study on the oxidation of same set of α-amino acids by pyridinium hydrobromide perbromide (PHPB) in aqueous acetic acid,[105] almost similar results were reported regarding the correlation between log k and σ^* and E_s (Eqs. 25–27).

$$\log k = -2.03(\pm 0.34)\sigma^* - 1.63$$
$$n = 9 \qquad r = 0.914 \qquad s = 0.22 \tag{25}$$

$$\log k = -0.74(\pm 0.05)E_s - 1.79$$
$$n = 9 \qquad r = 0.984 \qquad s = 0.097 \tag{26}$$

$$\log k = -0.71(\pm 0.01)\sigma^* - 0.54(\pm 0.01)E_s + \text{constt}$$
$$n = 9 \qquad r = 0.999 \qquad s = 0.004 \tag{27}$$

However, for the same series, the rate of oxidation by *N*-bromobenzamide was shown[106] to be correlated with σ^* and E_s as:

$$\log k = -0.82\sigma^* + 0.49E_s + \text{constt}$$
$$n = 9 \qquad r = 0.998 \qquad s = 0.014 \tag{28}$$

indicating a rate-retarding effect of steric factor. The negative steric effect here may be due to the hinderance produced by the substituent in the formation of substrate-oxidant complex that finally dissociates into the product. In each study, however, the resulting product was found to be corresponding aldehydes.

III. CONCLUDING REMARKS

In majority of the cases cited here, the value of the reaction constant (ρ, ρ^+, ρ^-, ρ^*, or ρ_R or ρ_F) is found to be negative suggesting that the electron-releasing substituents would usually accelerate and the electron-withdrawing substituents would usually retard the rate of oxidation reactions. The *ortho*-substituents may have some additional rate-retarding steric effects.

The vast majority of organic oxidations involve the rupture of carbon–hydrogen or carbon–carbon bonds. A carbon–hydrogen bond may be broken homolytically by hydrogen atom abstraction or heterolytically by either proton or hydride ion abstraction. Electron-releasing substituents can increase the electronic density at the reaction centres, say at carbon in carbon–hydrogen bond, facilitating the attack of electrophilic reagents that may result into the breaking of bonds. Also an increase in the electron density at the reaction centre may facilitate the flow of electrons from the substrates towards the oxidant. Acceptance by the oxidant of an electron is equivalent to the acceptance of a hydride ion. Thus, reactions can be classified as 1-equivalent or 2-equivalent reactions according to the transfer that occurs during the process. Organic compounds, with certain important exceptions, tend to avoid reactions requiring 1-equivalent steps because they are stable compounds and not radicals. However, in many-step reactions, a reaction intermediate can be in a radical form, and then it may require 1-equivalent step.

Since electronic charge separation between the initial and transition states is small in 1-equivalent reactions, the ρ value is usually low.[7–12] A high ρ value is associated with 2-equivalent reactions.[13–18] Thus, the high and low values of ρ can suggest whether the oxidation involves two-electron transfer or one-electron transfer, e.g. iodate or periodate oxidations of diphenylamines (**2**) and phenols with ρ values of the order of -6 and -5 indicate the two-electron transfer. Thus they provide clue to the mechanism of the reactions. In the sulfur-containing compounds, the marked negative values indicate that the sulfur atom of compound is highly positively charged, so that it strongly attracts the electron from electron-releasing substituents.

The sign of ρ value is of great significance. While a negative value suggests that electron-releasing substituents will enhance the rate of the oxidation reaction, a positive value indicates that the electron-withdrawing and not the electron-releasing substituents will enhance the rate of the reaction and thus suggests a different mechanism of the reaction. Though in most of the cases the ρ values are negative, in the case of benzaldehydes most of the oxidants, e.g. chromic acid,[50,55] Ni(III),[56] *N*-bromoacetamide,[58] quinolinium dichromate,[59] gave positive ρ values. Hence, it was assumed that the rate-controlling step involved the loss of H^+, which could be facilitated by increase in the positive charge at the carbonyl carbon by electron-withdrawing substituents. Thus, a heterolytic fission of C—H bond of aldehyde group was proposed in the oxidation of benzaldehyde. Otherwise, with the negative values obtained with some oxidants in the oxidation of benzaldehydes, a free-radical mechanism was proposed,[57] in which the rate-determining step was assumed to produce a free-radical ($Ar\dot{C}O$) which could be involved in the next step in the formation of a transition state. This free-radical mechanism was recently found to be valid even for the oxidation of aliphatic aldehydes (XCHO) with a value of $\rho^* = -2.5$ in a bis(dihydrogentellurato)cuprate(III) oxidation.[107] With a positive value of ρ equal to 0.85 in bis(dihydrogentellurate)cuprate(III) oxidation of benzaldehyde, however, the heterolytic fission of C—H bond (of aldehyde group) via proton transfer was restressed.[107]

However, in aromatic compounds like anils, where there are two phenyl rings, e.g. **4** and **7**, if Eq. 5 is found to hold good, the attack of oxidant on the double bond can be assumed to be a single concerted step.

Finally, it goes without saying that various reaction constants, i.e. ρ, ρ^+, ρ^-, etc., simply indicate the mode of transfer of electrons to, or their withdrawal from, the reaction centre by the substituents.

REFERENCES

1. Hammett, L. P. *Chem. Rev.* **1935**, *17*, 125.
2. Swain, C. G.; Lupton, Jr., E. C. *J. Am. Chem. Soc.* **1968**, *90*, 4328.
3. Hansch, C.; Leo, A.; Taft, R. W. *Chem. Rev.* **1991**, *91*, 165.
4. Taft, R. W. In *Steric Effects in Organic Chemistry*; Newman, M. S., Ed.; Wiley: New York, **1956**, p. 556.
5. Wiberg, K. B. *Oxidations in Organic Chemistry*; Part A, Academic Press: New York, **1965**.
6. Sheldon, R. A.; Kochi, T. K. *Metal-Catalyzed Oxidations of Organic Compounds*; Academic Press: New York, **1981**.
7. Bartlett, P. D.; Ruchard, C. *J. Am. Chem. Soc.* **1960**, *82*, 1756.
8. Watanabe, Y.; Iyanagi, T.; Oae, S. *Tetrahedron Lett.* **1980**, *21*, 3685.
9. Watanabe, Y.; Numata, T.; Iyanagi, T.; Oae, S. *Bull. Chem. Soc. Jpn.* **1981**, *54*, 1163.
10. Howard, J. A.; Ingold, K. U. *Can. J. Chem.* **1963**, *41*, 1744.
11. Russel, G. A.; Williamson, R. C., Jr. *J. Am. Chem. Soc.* **1964**, *86*, 2357.
12. Huang, R. L.; Lee, K. H. *J. Chem. Soc.* **1966**, 935.
13. Jaffe, H. H. *Chem. Rev.* **1953**, *53*, 191.
14. Brown, H. C.; Okamoto, Y. *J. Org. Chem.* **1957**, *22*, 485.
15. Brown, H. C.; Okamoto, Y. *J. Am. Chem. Soc.* **1958**, *80*, 4979.
16. Brown, H. C.; Bernheimer, R.; Kim, C. J.; Schepple, S. E. *J. Am. Chem. Soc.* **967**, *89*, 370.
17. Shiner, V. J., Jr.; Buddenbaum. W. E.; Murr, B. E.; Lamaty, G. *J. Am. Chem. Soc.* **1968**, *90*, 418.
18. Noyce, D. S.; Kaiser, G. V. *J. Org. Chem.* **1969**, *34*, 1008.

19. Leffler, J. E.; Grunwald, E. *Rates and Equilibria of Organic Reactions*; Wiley: New York, **1963**.
20. Shorter, J. *Correlation Analysis in Organic Chemistry. An Introduction to Linear Free-Energy Relationships*; Clarendon Press: New York, **1973**.
21. Johnson, C. D. *The Hammett Equation*; Cambridge University Press: New York, **1973**.
22. (a) *Advances in Linear Free-Energy Relationships*; Chapman, N. B.; Shorter, J., Eds.; Plenum: London, **1972**; (b) *Correlation Analysis in Chemistry*; Chapman, N. B.; Shorter, J., Eds.; Plenum: New York, **1978**.
23. Exner, O. In ref. 22(a), p. 37.
24. Sharadamani, P. R.; Jaganadham, V. *Indian J. Chem.* **1990**, *29A*, 700.
25. Vedavrath, R.; Sethuram, B.; Navaneeth Rao, T. *Proc. Natl. Acad. Sci. India* **1981**, *51A*, 443.
26. Prasada Rao, M.; Sethuram, B.; Navaneeth Rao, T. *Indian J. Chem.* **1979**, *17A*, 52.
27. Sreenivasulu, P. V.; Adinarayana, M.; Sethuraman, B.; Navaneeth Rao, T. *Oxid. Commun.* **1985**, *8*, 199.
28. Venkateshwarlu, S.; Jaganadham, V. *Indian J. Chem.* **1988**, *27A*, 314.
29. Venkateshwarlu, S.; Jaganadham, V. *Oxid. Commun.* **1985**, *8*, 149.
30. Srinivasan, C.; Perumal, S.; Arumugam, N. *Indian J. Chem.* **1980**, *19A*, 160.
31. Boyland, E.; Manson, D.; Sims, P. *J. Chem. Soc.* **1953**, 3623.
32. Boyland, E.; Manson, D.; Sims, P. *J. Chem. Soc.* **1954**, 980; *ibid* **1958**, 4198.
33. Behrman, E. J. *J. Am. Chem. Soc.* **1967**, *89*, 2424.
34. Srinivasan, C.; Perumal, S.; Arumugam, N. *J. Chem. Soc. Perkin Trans 2* **1985**, 1855.
35. Behrman, E. J.; Behrman, D. M. *J. Org. Chem.* **1978**, *43*, 4551.
36. Karunakaran, C.; Palaniswamy, P. N. *Gazz. Chim. Ital.* **1997**, *127*, 559.
37. Bramaiah, A.; Manikyamba, P.; Sundaram, E. V. *Indian J. Chem.* **1991**, *30A*, 247.
38. Manikyamba, P.; Ragunath Rao, P.; Sundaram E. V. *Indian J. Chem.* **1981**, *20A*, 1186.
39. Manikyamba, P.; Ragunath Rao, P.; Sundaram E. V. *Indian J. Chem.* **1981**, *20A*, 1186.
40. Radhakrishna Murthy, P. S.; Pati, S. C.; Janardhan, C. H. *Indian J. Chem.* **1978**, *16A*, 142.
41. Radhakrishna Murthy, P. S.; Pati, S. C.; Sri Ramulu, Y. *Indian J. Chem.* **1976**, *14A*, 955.
42. Vijaya Lakshmi and Sundaram, E. V. *Indian J. Chem.* **1977**, *15A*, 612.
43. Sharma, L. D.; Gupta, R. C.; Srivastava, S. P. *Indian J. Chem.* **1975**, *13*, 978.
44. Reddy, T. S.; Jagannadham, V. *Proc. Natl. Acad. Sci. India* **1986**, *56A(II)*, 129.
45. Manikyamba, P.; Ragunath Rao, P.; Sundaram E. V. *Indian J. Chem.* **1981**, *20A*, 1217.
46. Swain, C. G.; Langsdorf, W. P., Jr. *J. Am. Chem. Soc.* **1951**, *73*, 2813.
47. Ballistere, F. P.; Maccarone, E.; Momo, A. *J. Org. Chem.* **1976**, *41*, 261.
48. Rogne, O. *J. Chem. Soc. B* **1968**, 1294.
49. Panigrahi, G. P.; Misra, P. K. *Indian J. Chem.* **1977**, *15A*, 1066.
50. Musala Reddy, P.; Jagannadham, V.; Sethuram, B.; Navaneeth Rao, T. *React. Kinet. Catal. Lett.* **1982**, *21*, 437.
51. Lucchi, E. *Bull. Sci. Fac. Chim. Ind. Bologna* **1940**, *1*, 208, 333.
52. Lucchi, E. *Bull. Sci. Fac. Chim. Ind. Bologna* **1941**, *2*, 165, 176.
53. Lucchi, E. *Gazz. Chim. Ital.* **1941**, *71*, 729, 752.
54. Graham, G. T. E.; Westheimer, F. H. *J. Am. Chem. Soc.* **1958**, *80*, 3030.
55. Wiberg, K. B.; Mill, T. *J. Am. Chem. Soc.* **1958**, *80*, 3022.
56. Reddy, T. R.; Murthy, G. S. S.; Jagannadham, V. *Oxid. Commun.* **1986**, *9*, 83.
57. Murthy, G. S. S.; Sethuram, B.; Navaneeth Rao, T. *Indian J. Chem.* **1977**, *15A*, 1073.
58. Venkateshwarlu, S.; Jaganadham, V. *Oxid. Commun.* **1985/86**, *8*, 149.
59. Aruna, K.; Manikyamba, P.; Sundaram, E. V. *Asian J. Chem.* **1994**, *6*, 542.
60. Rindone, B.; Santaniello, E.; Scolastico, C. *Tetrahedron Lett.* **1972**, *1*, 19.
61. Catto, A.; Corboni, F.; Rindone, B.: Scolastico, C. *Tetrahedron Lett.* **1973**, *29*, 2723.

62. Narasimhabharathi, S.: Sundaram, S.; Venkatasubramanian, S. *Indian J. Chem.* **1977**, *15B*, 376.
63. Krishnamoorthy, N. Ph.D. Thesis. *A Kinetic and Mechanism Study of Bromate Oxidations*, Kakatiya University, India, **1986**.
64. Karunakaran, K.; Nagarajan, S.; Kanagvel, D.; Jegdish, T. N.; Palaniswamy P. N.; Elango, K. P. *Oxid. Commun.* **1997**, *20*, 576.
65. Ruasse, M. F.; Argile, A. *J. Org. Chem.* **1983**, *48*, 202.
66. Argile, A.; Ruasse, M. F. *J. Org. Chem.* **1983**, *48*, 209.
67. Narsimha Char, P.; Sondu, S.; Sethuram, B.; Navaneeth Rao, T. *Indian J. Chem.* **1988**, *27A*, 31.
68. Mohamed Kasim, A. N.; Basheer Ahamed, K. A. *Indian J. Chem.* **1999**, *38A*, 533.
69. Padmaja, S.; Nageshwar Rao, K.; Prasad Rao, M.; Sethuram, B. *Indian J. Chem.* **1995**, *34A*, 351.
70. Sondu, S.; Sethuram, B.; Navaneeth Rao, T. *Oxid. Commun.* **1987**, *10*, 167.
71. Srinivasan, C.; Kuthalingam, P.; Arumugam, N. *Can. J. Chem.* **1978**, *56*, 3043.
72. Rajasekaran, K.; Baskaran, T.; Gnanasekaran, C. *Indian J. Chem.* **1983**, *22A*, 1041.
73. Rajasekaran, K.; Baskaran, T.; Gnanasekaran, C. *J. Chem. Soc. Perkin Trans. 2* **1984**, 1183.
74. Srinivasan, C.; Perumal, S.; Arumugam, N. *J. Chem. Soc. Perkin Trans. 2* **1984**, 2065.
75. Srinivasan, C.; Perumal, S.; Arumugam, N. *Indian J. Chem.* **1985**, *24A*, 56.
76. Baliah, V.; Uma, M. *Tetrahedron* **1963**, *19*, 455.
77. Srinivasan, C.; Kuthalingam, P.; Arumugam, N. *J. Chem. Soc. Perkin Trans. 2* **1980**, 170.
78. Srinivasan, C.; Chellamani, A.; Kuthalingam, P. *J. Org. Chem.* **1982**, *47*, 428.
79. Perumal, S.; Srinivasan, C.; Arumugam, N. *Indian J. Chem.* **1986**, *25A*, 367.
80. Chellamani, A.; Alhaji, N. M. I. *Indian J. Chem.* **1999**, *38A*, 888.
81. Chellamani, A.; Alhaji, N. M. I.; Rajagopalan, S.; Sevvel, R.; Srinivasan, C. *Tetrahedron*, **1995**, *51*, 12677.
82. Chellamani, A.; Alhaji, N. M. I.; Rajagopalan, S. *J. Chem. Soc. Perkin Trans. 2* **1997**, 299.
83. Curci, R.; Modena, G. *Tetrahedron* **1963**, *19*, 1749; *Gazz. Chim. Ital.* **1964**, *94*, 1257.
84. Srinivasan, C.; Jegatheesan, P. P.; Arumugam, N. *Indian J. Chem.* **1986**, *25A*, 678.
85. Chellamani, A.; Kulanthaipandi, P.; Rajagopalan, S. *J. Org. Chem.* **1999**, *64*, 2232.
86. Ganapathy, V.; Jegatheesan, P. P. *Int. J. Chem. Kinet.* **1988**, *15*, 129.
87. Murray, R. W.; Jeyaraman, R.; Krishnapillai, M. *J. Org. Chem.* **1987**, **52**, 746.
88. Bloodworth, A. J.; Melvin, T.; Mitchell, J. C. *J. Org. Chem.* **1988**, *53*, 1078.
89. Suthakaran, R.; Subramanian, P.; Srinivasan, C. *Proc. Indian Acad. Sci. Chem. Sci.* **1986**, *97*, 555.
90. Acquaye, J. H.; Muller, J. G.; Takeuchi, K. *J. Inorg. Chem.* **1993**, *32*, 160.
91. Srinivasan, C.; Jegatheesan, P. P. *Indian J. Chem.* **1989**, *28A*, 250.
92. Srinivasan, C.; Venkatasamy, R.; Rajagopalan, S. *Indian J. Chem.* **1981**, *20A*, 505.
93. Srinivasan, C.; Jegatheesan, P. P.; Rajagopalan, S.; Arumugam, N. *Can. J. Chem.* **1987**, *65*, 2421.
94. Baliah, V.; Satyanarayana, P.V. V. *Indian J. Chem.* **1978**, *16A*, 966.
95. Cerniani, A.; Modena, G. *Gazz. Chim. Ital.* **1959**, *89*, 843.
96. Srinivasan, C.; Pitchumani, K. *Indian J. Chem.* **1979**, *17A*, 162.
97. Larsson, E. *Chem. Abstr.* **1949**, *43*, 8815h; **1952**, *46*, 9956f.
98. Srinivasan, C.; Subramanian, P.; Chellamani, A. *Indian J. Chem. Sci.* **1989**, *3*, 61.
99. Srinivasan, C.; Pitchumani, K. *Bull. Chem. Soc. Jpn.* **1982**, *55*, 289.
100. Karunakaran, C.; Gnanasekaran, K.; Alamelu, R. *AFINIDAD LII* **1995**, *458*, 245.
101. Srinivasan, C.; Subramanian, P. *J. Chem. Soc. Perkin Trans. 2* **1990**, 1061.

102. Thyagarajan, B. S. *Chem. Rev.* **1958**, *58*, 439.
103. Wilson, I. R. *Rev. Pure Appld. Chem.* **1966**, *16*, 103.
104. Pareek, A.; Kothari, S.; Banerji, K. K. *Indian J. Chem.* **1996**, *35A*, 116.
105. Devi, J.; Kothari, S.; Banerji, K. K. *Indian J. Chem.* **1995**, *34A*, 116.
106. Agarwal, A.; Mittal, S.; Banerji, K. K. *Indian J. Chem.* **1987**, *26A*, 339.
107. Sen Gupta, K. K.; Nandy, B. K.; Bera, A. K.; Sen Gupta, S. *Indian J. Chem.* **1997**, *36A*, 190.

MOLECULAR GEOMETRY AS A SOURCE OF CHEMICAL INFORMATION—APPLICATIONS OF THE BOND VALENCE–BOND NUMBER MODELS

Sławomir J. Grabowski and Tadeusz M. Krygowski

Advances in Quantitative Structure Property Relationships
Volume 3, pages 27–66.

I. INTRODUCTION

Bond number[1] and bond valence[2] concepts originate from the very fundamental concept of the valence as a property of atoms of elements to be bonded with other atoms. Even if the modern chemistry[3] does not use valency of elements as an advanced idea, it is still a very useful tool in many basic interpretations. In parallel with the disappearence of the idea of valence in modern chemistry, there has appeared an equivalent theoretical concept termed a bond order.[4] This concept has also been formulated within empirical models of structural chemistry.[5,6] In principle all these concepts: bond order, bond number and bond valence represent approximately the same feature—an amount of electron cloud between nuclei which are bound in some kind of a fixed structure: molecules or crystals.

In this chapter we intend to show some part of these ideas, which are limited to the formulation and applications of Pauling's[1] and Brown's[2] ideas of bond number and bond valence, respectively.

II. THE DEFINITION OF THE BOND NUMBER

In 1947 for the interatomic distances observed in metals Pauling introduced the concept of bond number n, related with distances by the equation[1]:

$$r_{\mathrm{n}} - r_{\mathrm{s}} = \Delta \mathrm{r} = -\mathrm{c} \log n \tag{1}$$

where r_{n} and r_{S} designate bond lengths, the given one with a bond number equal to n, and a single one treated as a reference with bond number equal to 1. c is the constant which varies depending on the nature of atoms involved in the bond. The c value may be determined for the appropriate pair of connected atoms if the value of n is known for the corresponding distance between atoms. Apart of the original use of the concept of bond number for metals, it may be applied either to non-covalent interactions or to covalent bonds. In both cases it is necessary to calculate (estimate) value of c in Eq. 1. In the case of non-covalent interactions, the r_{n}-value is taken most often from the symmetrical systems. For example, for O—H...O bridges, the system with the centric position of proton is taken into account. Then, for such system we have two equivalent half-bonds; and Eq. 1. leads to the numerical value of c:

$$r_{1/2} - r_s = -c \log \frac{1}{2} \tag{2}$$

For the symmetrical O—H...O system the proton lies in the middle of the O...O distance equal to 2.44 Å,[7] hence we have the length of half-bond equal to 1.22 Å. However in the last investigations on the resonance assisted H-bonds[8] the O...O distance of 2.4 Å is proposed as the shortest one with the proton in the center. It appears here that the choice of the reference interatomic distance must be arbitrary, which is an obvious bias of most empirical approaches. It should be mentioned here, that r_S value corresponds in Pauling's approach to the single bond, but in some of the investigations this value is treated as the usual constant fitting some set of experimental data, and may be different from the typical length of the single bond.[9]

For the further discussion we have to answer the question: what is the physical meaning of the bond number? In his work on interatomic distances in metals Pauling defined[1] the quantity n as the number (or better a fraction) of electron pairs participating in the atom–atom contact or in a bond between atoms. Additionally the logarithmic relation defining the bond number (Eq. 1.) is a consequence of the exponential character of intermolecular forces. Hence n is often considered almost equivalently as "bond order", "bond valence" or "bond strength".

The relationship between the bond number n and the value of bond length depends on the nature of the bond. It is shown in Figure 1 for pairs of interacting atoms: O—H (H...O), N—H (H...N), C—H (H...C) and F—H (H...F). It is worth mentioning the sequence of varied bond length in increasing order (F...H, O...H, N...H, C...H) for a fixed bond number. This is in line with a decrease of the difference of electronegativity of atoms forming the bond (or the interatomic contact). Consequently, the elongation of X—H bond is observed.

Figure 1 confirms the previously presented statement, that the equation (1) may be applied to both, the typical bonds as well as to the intermolecular interactions. It is evidently shown by the range of variations on abscissa where bond length (interatomic distance) is in the range between 0.9 and 2.1 Å.

Apart of a wide use of the bond number concept to H-bonded systems, this idea was also applied[10,11] to triatomic systems like $(Br...Br...Br)^-$, $(I...I...I)^-$, $(Cl...I...Cl)^-$ etc. Table 1 presents the constants used in Eq. 1 for the above mentioned systems.

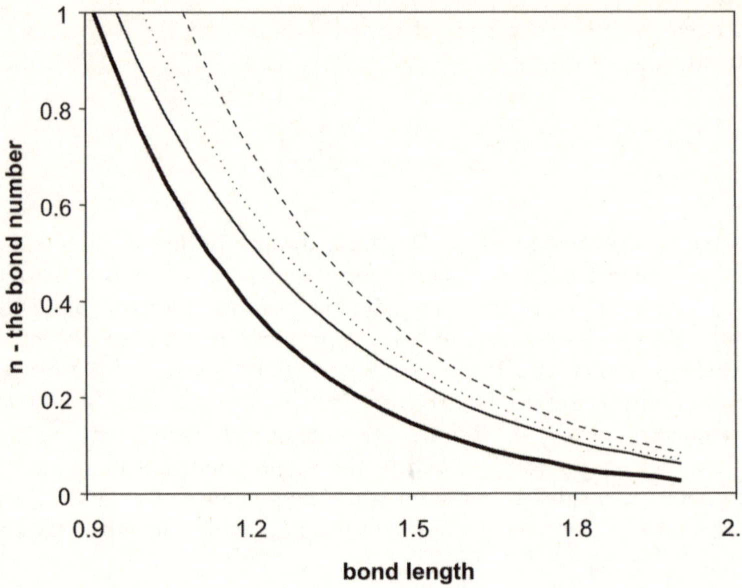

Figure 1. The dependence between the bond length and its bond number (Eq. 1); solid bold line correspond to F—H, solid usual line to O—H, dotted line to N—H and broken line to C—H.

In the interpretation of the results for triatomic systems Bürgi[10] and Dunitz[11] adopted Pauling's relationship (Eq. 1) with the additional assumption, postulating that the sum of two bond numbers within the system considered equals to unity which leads to the equation:

$$10^{-\Delta r1/c} + 10^{-\Delta r2/c} = 1 \tag{3}$$

where $\Delta r1 = r1 - r_S$ and $\Delta r2 = r2 - r_S$ are related bond distance increments and correspond to $r1$ and $r2$ bond lengths within the system. The relation of the bond number conservation was first introduced by Johnston[6,14] to describe the gas phase reactions, originally it was called as bond order conservation rule and this relation expresses Pauling's postulate according to which the sum of all bonds around the central atom is constant.[1] This relationship will be described in detail in the next sections.

It is important to stress that the idea of bond number may be applied to any pair of the connected atoms: they may interact by typical intermolecular forces (H-bonding, van der Waals forces or others) or by typical covalent interactions. Thus the bond number idea may be applied to the fractional bonds but also to the double or triple bonds. Bonds with delocalized

Table 1. Values of r_S, $r_{1/2}$ and c (in Å) for various H-bonds and the other triatomic systems (mainly taken from ref. 11)

XHX	r_S (Å)	$r_{1/2}$ (Å)	c
O—H...O	0.957	1.22	0.87
$(F...H...F)^-$	0.917	1.13	0.70
N—H...N[a]	1.010	1.27	0.86
$(Br...Br...Br)^-$	2.284	2.54	0.86
$(I...I...I)^-$	2.667	2.93	0.86
$Cl...Br...Cl)^-$	2.138	2.40	0.86
$(Cl...I...Cl)^-$	2.321	2.55	0.76
$(Br...I...Br)^-$	2.485	2.71	0.76
S...S...S	2.02	2.34	1.03
H...H...H	0.741	0.93	0.62
C—H...C[d]	1.08[b]	1.34	0.85[c]

Notes: [a] data taken from ref. 12.
[b] the value proposed by Jeffrey and Lewis[13] as the most appropriate for single C—H bond length.
[c] the constant value proposed by Brown and Altermatt[9] as typical for different kinds of covalent bonds described by Eq. 1 and applied here for C—H bonds (the authors used a different expression of Eq. 1, so in this work the data given in ref. 9 were recalculated).
[d] $n_{1/2}$ value for C—H...C systems (1.34 Å) was calculated from Eq. 1 applying the given constants (1.08 and 0.85)

electrons may also be considered (for example the π-electron systems[15,16]). In the latest cases the postulate of conservation of bond number may be also applied, leading to a possibility of the description of a given atom and its ability for various kinds of reactivity on it.

However it should be stressed that Eq. 1 should not be applied for too long range of interatomic distances,[9] since the model is empirical and too long extrapolation may lead to confusion.

III. THE BOND–VALENCE MODEL

Another approach relating the bond length to the expression describing the amount of the electron clouds along the bond (or the path of interactions) comes from Brown.[2,17] He has also analyzed the relationships between the bond lengths and their physical properties and energy parameters. Brown formulated the bond–valence model (BV model) based on the following assumptions. Bonds are characterized by their bond valences (s) and their bond lengths. The relationship between bond length and valence bond is monotonic and may be approximated by simple functions:

$$S_{ij} = \exp\left[\frac{(r_S - r_{ij})}{B}\right] \quad \text{or,} \tag{4}$$

$$S_{ij} = \left(\frac{r_{ij}}{r_S}\right)^{-N} \tag{5}$$

where r_S, B and N are fitted constants.

Equation 4 may be easily transformed into the Eq. 1 defining the Pauling bond number n. Then one can see that the bond number and the valence bond describe the same or alike quantities what was pointed out by Brown.[2] Equation 4 is the one most often applied for the investigations concerning correlations between structural parameters.[2,8,9,12,16–18] As it was mentioned earlier r_S is the fitted constant not always equal to the typical single bond length in spite of the fact that r_S within BV model is designated as the length of a bond of unit valence. Using the Inorganic Crystal Structure Database[19,20] the parameters needed to determine bond valences from bond lengths have been calculated for 750 atom pairs[9] and resulted in the data for 141 pairs. Table 2 presents some selected r_S values which may be used in Eq. 4 for the calculation of valence bonds.

It is clear that r_S values in Brown's model are different from the typical single bonds: O—H or N—H which are 0.95–0.96 Å (the gas electron diffraction values)[21] and 1.016 Å.[22] It results from the statistical approach in estimating r_S values. It is known, that O—H bond lengths in water molecules in hydrates (neutron diffraction data) vary significantly: 37% of the sample has O—H bond length shorter than the gas phase value.[23] Thus the physical meaning of this constant in the Brown approach is not well defined and depend on the sample taken into estimation.

B value has been set to 0.37 Å for all pairs of atoms. According to the assumptions of BV model Eqs. 4 and 5 are justified over the small range of atom–atom distances. Additionally atomic valences have been introduced within the model; bond valences are related to atomic valences through the equations:

$$\sum_j s_{ij} = V_i \tag{6}$$

$$\sum_{\text{loop}} s_{ij} = 0 \tag{7}$$

where i and j correspond to different atoms. Equation (6) was called as the valence-sum rule (VSR) and its applications in chemistry and physics will be described more in details in the next sections. VSR is similar to the bond number conservation rule (Eq. 3) described in the previous section. In fact

Table 2. Values of r_S's for the selected pairs of atoms (ref. 9), B has been set to 0.37 for all pairs, oxidation states of atoms are given in parantheses

Pair of atoms	r_S*-value (in Å)*
C(IV)...N(III)	1.442
C(IV)...O(II)	1.390
D(I)[a]...O(II)	0.927
H(I)...N(III)	0.885
H(I)...O(II)	0.882
N(III)...O(II)	1.361
N(V)...O(II)	1.432
P(V)...N(III)	1.704
P(V)...O(II)	1.617
P(V)...S(II)	2.145
S(II)...N(II)	1.597
S(II)...N(III)	1.682
S(IV)...N(III)	1.762
S(IV)...O(II)	1.644
S(VI)...O(II)	1.624
Si(IV)...C(IV)	1.883
Si(IV)...N(III)	1.724
Si(IV)...O(II)	1.624
Si(IV)...S(II)	2.126

Note: [a] D designates deuterium

the bond number conservation rule is a particular case of VSR and was applied to triatomic systems. V_i is the atomic valence which is defined to be equal to the formal oxidation state.

According to Eq. 7 the sum of the bond valences around any loop is zero. The sign of the valence bond corresponds to the direction in which the bond is traversed in completing the circuit.[17] Equations 6 and 7 are called the Network Equations and allow to predict bond lengths within the considered crystal structure. The different ways of the solution of the Network Equations were proposed[24–26] but the detailed analysis of these approaches will not be the subject of the present study.

The presented relationships and models show that there are different definitions and names for the parameters of a very similar meaning; bond number—bond valence—bond order, the valence sum rule—the bond number conservation rule. Hence in the next sections the original terminology of Pauling will be used, i.e bond number and the bond number conservation rule.

IV. THE DERIVATION OF THE BOND NUMBER FROM DIFFERENT POTENTIALS

A. Different Potentials–Different Definitions

Pauling explained the expression 1 defining the bond number as being the consequence of the exponential character of atom–atom bonding interactions. Hence the bond number could be related to different functions describing the interactions. Brown and Shannon[27] derived the definition of the bond number (originally named bond valence) from the Born–Landé potential:

$$U_{ij} = \frac{Aq_iq_j}{R_{ij}} + \frac{B}{R_{ij}^{\mathrm{N}}} \tag{8}$$

where; U_{ij} describes the energy of interactions between ith and jth atoms, q_i and q_j are the net atomic charges on these atoms, R_{ij} is the atom–atom distance and A, B are constants corresponding to the appropriate type of atoms.

For the equilibrium value of R_{ij} the first derivative of Eq. 8 equals to zero and hence:

$$\frac{2Aq_iq_j}{NB} = R_{ij}^{-(N-1)} \tag{9}$$

Following the arguments given by Brown and Shannon[27] we can assume that B is proportional to the coordination number, $n = K\,B$ and then we can rewrite Eq. 9 as:

$$\frac{q_i}{n} = \left(\frac{1}{2}AKq_j\right)R_{ij}^{-(N-1)} \tag{10}$$

where q_i/n is equal to V_{i}/n which is the expectation value of the bond number.

Another approach was suggested by Dunitz[11] who has pointed out that it is possible to derive the definition of the bond number applying the Morse function:

$$V = D_0\{\exp[-2B(R - R_0)] - 2n\exp[-B(R - R_0)]\} \tag{11}$$

R and R_0 are the atom–atom distances, of the given one and for the equilibrium state respectively. D_0 is the dissociation energy of the bond since the Morse function is most often applied for the description of covalent bonds. B is the constant value appropriate for the pair of atoms under consideration.

The attractive part of the Morse function was multiplied by the factor n corresponding to the bond number. It is possible to calculate the first derivative and to obtain from Eq. 11:

$$V' = D_0(-2B)\exp[-2B(R - R_0)] + 2nBD_0\exp[-B(R - R_0)] = 0 \qquad (12)$$

For the state of equilibrium the value of V' equals to zero. Equation 12 directly leads to the definition of the bond number:

$$n = \exp[-B(R - R_0)] \qquad (13)$$

Bürgi and Dunitz[28] have proposed to multiply the attractive term of the Morse function more generally by some power of the bond number—n^q what leads to the equation:

$$n = \exp\left[\frac{-B}{q}(R - R_0)\right] = \exp[-B'(R - R_0)] \qquad (14)$$

Such attempt seems to be more proper since the constant B of the Morse function do not have to be the same as the constant of the definition of the bond number. The general inverse power potential may be modified in the same way as the Morse function:[28]

$$V(R) = \left[\frac{D_0}{(1-m)}\left[m\left(\frac{R_0}{R}\right)^1 - n^q 1\left(\frac{R_0}{R}\right)^m\right]\right], \quad 1 > m \qquad (15)$$

leading to the equation:

$$n = \left(\frac{R}{R_0}\right)^{(m-1)/n} \qquad (16)$$

which corresponds to the Eq. 5, that also is used as the definition of the bond number (originally bond valence). Other types of interatomic potential functions may be modified in the similar way by multiplying the attractive terms of potentials by the n-bond number values.[29] For example, the Varshni potential is as follows:

$$V = D\left\{1 - \left(\frac{R_0}{R_i}\right)\exp\left[-\beta\left(R_i^2 - R_0^2\right)\right]\right\}^2 \qquad (17)$$

where D is the dissociation energy, R_i the bond length and β is the constant. It has been also pointed out that the Buckingham type of the function:

$$V = A\exp(-\alpha R_i) - \frac{B}{R_i^6} \qquad (18)$$

and the Rydberg potential:

$$V = D[1 + \alpha(R_i - R_0)]\exp[-\alpha(R_i - R_0)] \quad (19)$$

may lead to the definitions of the bond number.

R_i in Eqs. 17–19 is the bond length and A, B, D, α, β and R_0 are constants. Three different functions corresponding to the definitions of the bond number may be obtained from Eqs. 17–19:

$$n_V = \frac{a}{R}\exp(-\beta R^2),$$
$$n_B = CR^7\exp(-\alpha R),$$
$$n_R = \frac{1}{(b + gR)}. \quad (20)$$

R instead of R_i is used in Eqs. 17–19 indicating the distance between atoms. The subscript corresponds to the type of the function from which the definition was obtained (V–Varshni, B–Buckingham, R–Rydberg). Each of these definitions (Eqs. 20) contain two constants. The way of the calculation of these constants was described earlier. It may be done from the definition of n (bond number) applied for single and for fractional bond lengths. It is worth mentioning that Eq. 14 obtained from the Morse function contains also two constants: B–value and R_0. Equations 20 may be modified leading to the similar dependences as Eq. 14: one general constant and the second one—R_0 corresponding to the length of the single bond.

Figure 2 illustrates the above presented discussion on the application of differently defined potentials to obtain the function dependence of the bond number on the bond length. There is shown a relationship between n-value and the bond length for the mentioned above definitions (Eq. 14 and Eqs. 20) of bond number. The O—H bond length was taken into account here as an example of such dependence.

We see that only the n-values obtained from the Rydberg-type definition differ significantly from the values obtained from the other definitions. The range of O—H lengths corresponds not only to typical covalent bonds but also to the H...O intermolecular contacts.

B. The Dependence Between O...O Distance and O—H Bond Length

The bond number conservation rule (Eq. 3 and more general the VSR—Eq. 6) applied to O–H...O system has got the form:

$$n_{\text{O–H}} + n_{\text{H...O}} = 1 \quad (21)$$

The different definitions of the bond number may be used within Eq. 21.

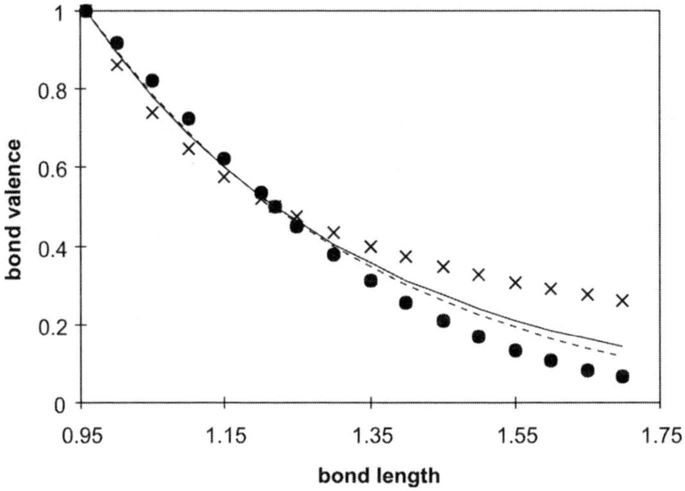

Figure 2. The dependence of the bond length (in Å) on the bond number obtained from different atom–atom potentials. The solid line corresponds to the Morse potential from which the definition of the bond number was obtained, the broken line corresponds to the Varshni potential, full circles to the Buckingham potential and the crosses to the Rydberg function.

Hence it is possible to obtain different dependences between the O—H bond length and the H . . . O distance for O—H . . . O system.

$$
\begin{gathered}
\mathrm{A}\exp(-\mathrm{B}\ \mathrm{R_{OH}}) + \exp(-\mathrm{B}\ \mathrm{R_{H...O}}) = 1 \\
\frac{a}{R_{\mathrm{OH}}}\exp(-\beta R^2_{\mathrm{OH}}) + \frac{a}{R_{H...O}}\exp(-\beta R^2_{\mathrm{H...O}}) = 1 \\
CR^7_{\mathrm{OH}}\exp(-\alpha R_{\mathrm{OH}}) + CR^7_{\mathrm{H...O}}\exp(-\alpha R_{\mathrm{H...O}}) = 1 \\
\frac{1}{(b + gR_{\mathrm{OH}})} + \frac{1}{(b + gR_{\mathrm{H...O}})} = 1
\end{gathered}
\tag{22}
$$

R_{OH} designates the length of O—H bond, $R_{\mathrm{H...O}}$ is the H . . . O distance, the remaining correspond to the constants. Equation 22 shows simply the relations between O—H bond lengths and H . . . O distances and hence for the linear O—H . . . O systems, they are relations between O—H length and O . . . O distance. Such relationships are present in Figure 3.

The designations of Figure 3 correspond to those of Figure 2. One can see that the relationship based on the Rydberg-type functions (crosses) again differs most significantly from the remaining curves. The relationship based on the Buckingham potential (circles) is practically the same as that obtained from the Morse-type function (solid line). Figure 3 also presents

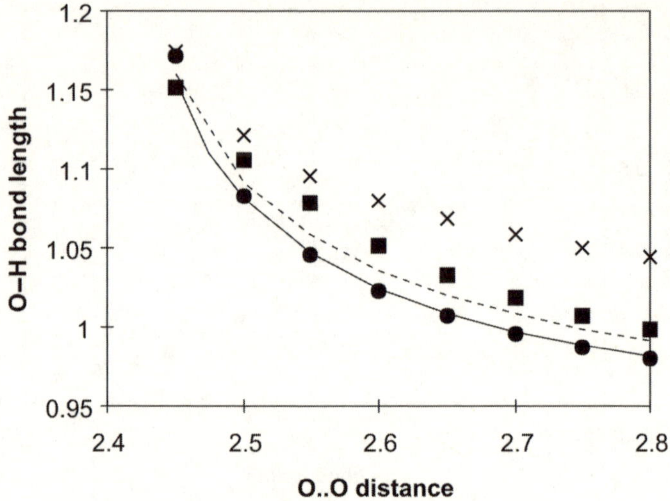

Figure 3. The dependences between O...O distances and O—H bond lengths obtained from Eqs. 22.

the same relation obtained from the Lippincott–Schroeder model[30,31] (LS-model) of hydrogen bonding (squares).

The relationships of Figure 3 were obtained from the bond number conservation rule (BNCR) where bond number definitions were derived from different potentials.[29]

C. The Dependence Between H...O Distance and O—H Bond Length

The dependence between H...O distance and O—H bond length derived from BNCR was compared with the high precision results of neutron diffraction[8,16,32] (Figure 4).

Solid line of Figure 4 represents the relationship obtained from BNCR (Eqs. 21 and 22) applying the definition of bond number derived from the Morse-like function; the constant values applied here slightly differ from those proposed by Dunitz for O—H...O systems, the bond length of a single O—H bond is the same, 0.957 Å, but the $r_{1/2}$ from which the constant c of Eq. 1 or constant B of Eq. 4 may be obtained is equal to 1.2 Å as corresponding to the shortest known O...O distance, 2.4 Å, for centric position of H-atom.

Broken line represents the BNCR relation proposed by Gilli and coworkers for O—H...O systems.[8] The r_0 value of 0.925 Å was put here instead as the reference single O—H bond length. The authors claim[8] that because of the nature of interactions within O—H...O systems the new r_0

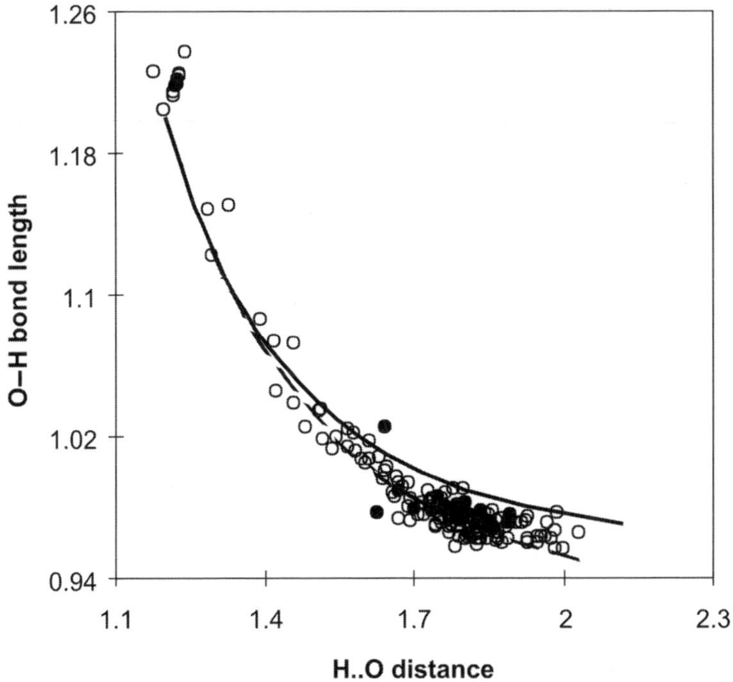

Figure 4. O—H bond length versus H...O distance for the set of accurate neutron diffraction results on 144 O—H...O bonds (open circles), O—D...O bonds are also included (full circles). Solid line represents the function obtained from the bond number conservation rule; distances in Å (figure taken from ref. 32).

value better represents the changes of geometries of H-bridges. The disagreement between the continuous line (the original BNCR relation with r_0 equal to 0.957 Å) and the neutron diffraction data is greater for longer H...O distances because of the the greater contribution of the electrostatic interactions.[8]

The proposal of Pauling concerning the bond number–bond length relationship and the proposal of Dunitz concerning the constant values for O—H...O systems are justified by their physical meanings, the exponential function reflecting the exponential nature of interactions and the reference single bond for the bond number equal to unity. The approach presented by Gilli[8] has the physical meaning and it is also justified by the type of interactions. However, in a lot of works the constant values of the definition of the bond number are fitted to the appropriate experimental data. Such approach was presented here within the bond valence model.[9] A similar fitting was demonstrated by Steiner[18] for the neutron diffraction data on H-bridges. Table 3 shows constant values for different sets of neutron

Table 3. The constant values appropriate to calculate bond numbers for O—H bonds and H...O contacts for different O—H...O neutron diffraction systems. The part of values given in the table was recalculated from the constants applied for Eq. 4 into the constants of Eq. 1.

H-bond type	r_S (Å)	$r_{1/2}$(Å)	*c*(Å)	*Ref.*
O—H...O, all types, LT[a]	0.928	1.2	0.90	[33]
O—H...O, all types	0.925	1.2	0.91	[8,34]
O_{water}—H...O	0.911	1.19	0.93	[35]
$O_{hydroxy}$—H...O	0.934	1.2	0.89	[18]
O—H...O[b]	0.957	1.22	0.87	[11]

Notes: [a] LT—low temperature, the measurements for temperatures less than 150 K.
[b] the proposal of Dunitz.[11]

diffraction O—H...O systems given by Steiner[18] and their comparison with the other proposals concerning such constants.

It should be noted that such considerations as that for O—H...O interactions may be carried out for other contacts but they still need some additional work.

D. N—H...N Bonds

Figure 5 shows the dependence between H...N distance and N—H bond length for N—H...N bonds taken from accurate neutron diffraction data of Cambridge Structural Database (circles). The solid line represents the relationship obtained from BNCR using constants of Table 1 appropriate for N—H...N systems ($r_0 = 1.01$ Å; $r_{1/2} = 1.27$ Å and $c = 0.86$ Å). The broken line was obtained after fitting the curve resulting from BNCR to the neutron diffraction set of N—H...N systems[18] (different from this one presented here). The constant values for the second curve are: $r_0 = 0.996$ Å; $r_{1/2} = 1.26$ Å and $c = 0.88$ Å.

The broken line is in a much better agreement with the experimental data than the solid line. However, it is worth mentioning that it was obtained from the fitting to the similar set of neutron diffraction data as that presented here while the first relation (solid line) is based on the arbitrarily accepted and physically justified constants.

It may be concluded that slightly different forms of bond number/bond valences type quantities are equivalent if used in comperative studies, since they all rely in a monotonic way on bond lengths (or interatomic distances). Hence, independently of the definition used, the applications may be successful in any qualitative interpretation of the molecular geometry.

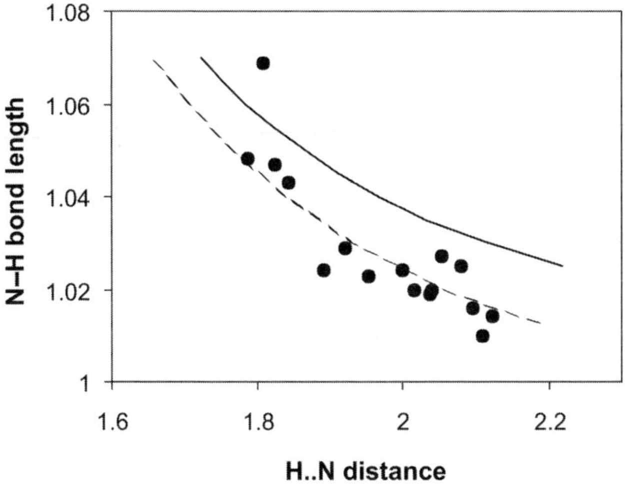

Figure 5. The relationship between H . . . N distance and N—H bond length for N—H . . . N bonds. Circles correspond to neutron diffraction results of the high precision considered in ref. 12; continuous and broken lines were obtained from BNCR.

V. ATOMS IN MOLECULES THEORY AND THE BOND NUMBER CONCEPT

A. Critical Points

The term "bond order" has a specific meaning in molecular orbital theory and there is another different meaning in valence bond theory (see ref. 4). As it was pointed out in the previous sections of this review, the term "bond number" or "bond valence" is rather connected with the empirical expotential or power type relation (Eqs. 4 and 5) most often connected with the hydrogen bonds or the other intermolecular interactions though the covalent bonds may be also considered.

A very interesting approach to the interpretation of the Pauling's bond number concept was demonstrated by Bader,[36] who has shown that the charge density can be used to define the bond order (which is closely related to the Pauling concept). The knowledge of the electron density, $\rho(r)$, is sufficient to describe the bonding pattern of any molecule. A useful tool to consider electron density in a molecule are electron density parameters in the critical points. They describe well the nature of the bond and are often used for this purpose. By definition, for the critical points the gradient of $\rho(r)$

vanishes:

$$\nabla\rho(r)_{r=r\,\mathrm{crit}} = 0 \tag{23}$$

Critical points are classified according to the number of negative eigenvalues of the Hessian of ρ (matrix of partial second derivatives with respect to $\{x,y,z\}$):

$$h_{\mathrm{pq}} = \frac{\partial^2}{\partial p \partial q}\rho(r)_{r=r\,\mathrm{crit}} \quad p, q = x, y, z \tag{24}$$

Stable critical points fall into one of four categories: maxima in $\rho(r)$ correspond to attractors (which are attributed to the nuclei), minima are connected with cage points and saddle points; the last mentioned ones correspond to the bond or ring points.[37,38] In other words they characterize charge density in the bond at the border between one and another atoms and in the ring, somewhere close to its center (if the ring is not the regular polygon).

B. The Bond Order as a Function of Electron Density

The definition of the bond order, n_{AB}, is connected with the electron density at the corresponding bond critical point—$\rho_{\mathrm{AB}}^{\mathrm{crit}}$:[36]

$$n_{\mathrm{AB}} = a_{\mathrm{AB}} \exp\left(b_{\mathrm{AB}}\ \rho_{\mathrm{AB}}^{\mathrm{crit}}\right) \tag{25}$$

The constants a and b of Eq. 25. depend on the nature of the bonding atoms A and B. Such definition requires parametrization which introduces arbitrariness in the choice of the reference molecules for which the reference bond orders are known. Furthermore Eq. 25 transformed into the logarithmic form may be related to the original Pauling equation, Eq. 1, in which bond lengths are replaced by the electron densities in the critical points of the bonds in question. Equation 25 may be also directly related to the exponential formula for the definition of bond valence (Eq. 5) leading for both cases to the linear relation between bond length and electron density in the critical point of the bond.

It is well known, that there is a very good linear dependence between the bond lengths and electron densities at bond critical points. For 58 data points for C—C bonds in benzenoid hydrocarbons, correlation is very high, $r = -0.998$.[39]

Alkorta et al.[40] proposed to replace the bond order, which does not have a unique meaning, with a computable parameter, i.e. the charge density at the bond critical point (ρ^{crit}). It was pointed out that the relationship between A-atom–B-atom distance and $\rho_{\mathrm{AB}}^{\mathrm{crit}}$ for the greater range of atom–atom distances, the covalent and hydrogen bond length is not linear but logarithmic.[40] Equations 26–28 show the logarithmic dependences obtained from B3LYP/6–311++G(d,p) calculations by Alkorta and coworkers[40] for C...H, N...H and F...H pairs of atoms (covalent bonds and intermolecular interactions):

$$r_{\mathrm{CH/H...C}} = (0.52 \pm 0.02) - (0.425 \pm 0.003)\ln(\rho_{\mathrm{bcp}}) \qquad n = 33 \qquad r^2 = 0.996 \tag{26}$$

$$r_{\mathrm{FH/H...F}} = (0.48 \pm 0.02) - (0.395 \pm 0.006)\ln(\rho_{\mathrm{bcp}}) \qquad n = 22 \qquad r^2 = 0.995 \tag{27}$$

$$r_{\mathrm{NH/H...N}} = (1.25 \pm 0.05) - (2.36 \pm 0.03)\ln(\rho_{\mathrm{bcp}}) \qquad n = 18 \qquad r^2 = 0.997 \tag{28}$$

Equations 27 and 28 were obtained from the calculations on $[\mathrm{F{-}H...F}]^-$ and $[\mathrm{H_3N{-}H...NH_3}]^+$ charged complexes. The relationships between ρ_{bcp}'s and appropriate bond lengths (Eqs. 26–27) are presented in Figure 6.

It was demonstrated that the bond number conservation rule applied to hydrogen bonding system may be expressed in terms of the values of charge densities at bond critical points.[41] The bond number conservation rule for the X—H...B system may be written as:

$$\exp\left[\frac{(R^0_{\mathrm{HX}} - R_{\mathrm{HX}})}{b}\right] + \exp\left[\frac{(R^0_{\mathrm{HB+}} - R_{\mathrm{H...B}})}{b}\right] = 1 \tag{29}$$

where R^0_{HX} and $R^0_{\mathrm{HB+}}$ correspond to the free XH and HB^+ bonds; R_{HX} and $R_{\mathrm{H...B}}$ are the appropriate distances within X—H...B hydrogen bonding. For the same kind of bond or atom–atom contact, one may find the relation of the type (as for example Eqs. 26–28):

$$R = A + B\ln\rho, \tag{30}$$

and for the difference;

$$\Delta R_i = R^0_{\mathrm{HX}} - R_{\mathrm{HX}}, \tag{31}$$

then we can write;

$$\Delta R_i = B[\ln\rho(R^0_{\mathrm{HX}}) - \ln\rho(R_{\mathrm{HX}})] \tag{32}$$

A similar relation may be written for the difference between HB^+ bond

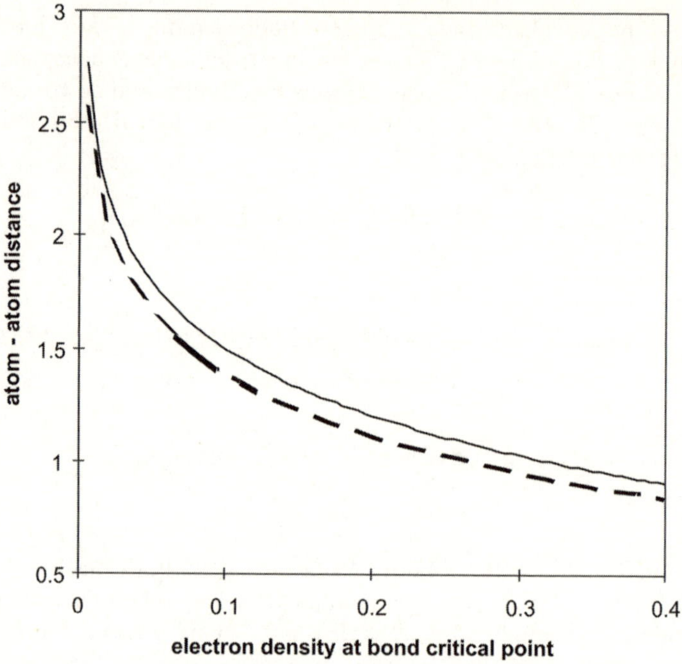

Figure 6. Relationship between electron density at the bond critical point and atom–atom distance; continuous line represents C and H pair of atoms, broken line is for F and H atoms, both relations were obtained from the data of ref. 40.

length and the H . . . B distance:

$$\Delta R_j = R^0_{\mathrm{HB+}} - R_{\mathrm{H...B}} = B'[\ln \rho(R^0_{\mathrm{HB+}}) - \ln \rho(R_{\mathrm{H...B}})] \tag{33}$$

Finally, the bond number conservation rule, in terms of charge densities at bond critical points, has the form:

$$\left[\frac{\rho(R^0_{\mathrm{HX}})}{\rho(R_{\mathrm{HX}})}\right]^{B/b} + \left[\frac{\rho(R^0_{\mathrm{HB+}})}{\rho(R_{\mathrm{H...B}})}\right]^{B'/b} = 1 \tag{34}$$

where b is the same constant as that within the definition of the bond number (c from Eq. 1 or B from Eq. 4).

It was demonstrated that the electron density at the bond critical point has additive properties for different hydrogen bonding systems.[41]

VI. BOND ENERGY–BOND ORDER METHOD (BEBO)

The bond-energy–bond-order (BEBO) method was worked up by Johnston and Parr.[6,14] It was developed and applied for the reactions of the transfer of a hydrogen atom:

$$\text{R—H} + \text{X} \rightarrow \text{R} \ldots \text{H} \ldots \text{X} \rightarrow \text{R} + \text{H—X}$$

Hence the relations connected with the Pauling's bond number are useful within the BEBO model; the bond number conservation rule applied for the mentioned above R—H...X system (Eqs. 3 and 21), the definition of the bond number (Eq. 1 or 4) and the relationship between the bond number and its energy:

$$D_n = D_S n^p \tag{35}$$

where D_n is the energy of the bond which may be described by the n-bond number, D_S is the bond energy of the single AB bond and p is an empirical parameter. The energies of R—H bonds and H...X interactions may be calculated from Eq. 35 if we know the bond numbers—n_{RH} and $n_{H\ldots X}$ respectively (D_S and p are constant values which may be designated for each type of pair of atoms). However for the R—H...X hydrogen bonding the R...X interaction is also of the great importance. For example, the most general semiempirical model of hydrogen bonding, proposed by Lippincott and Schroeder[30,31] takes into account the following terms within the H-bond energy (E_{HB}):

$$E_{HB} = E_{RH} + E_{H\ldots B} + E_{R\ldots B} \tag{36}$$

the terms concerning R—H bond, H...B contact and R...B interaction, respectively.

Johnston and Parr[14] within the BEBO method proposed for the R...B interaction ($E_{R\ldots B}$) the following relation:

$$E_{R\ldots B} = D_e\left[\left(\frac{1}{2}\right)\exp(-\beta R_{R\ldots B})\right]\left[1 + \left(\frac{1}{2}\right)\exp(-\beta R_{R\ldots B})\right] \tag{37}$$

originally called as a triplet anti-Morse function.

The last equation can be expressed in terms of bond order, since;

$$R_{R\ldots B} = r_{R\ldots B} - r^0_{R\ldots B} \tag{38}$$

$r_{R\ldots B}$ is the function of R—H bond length and H...B distance which may be replaced by the corresponding relations of the bond numbers (Eq. 4).

Hence it is possible to construct the final expression on H-bond energy—E_{HB} (Eq. 36).[14]

The explanation of the triplet interaction is needed. Johnston and Parr explained R...B interaction in the following way; if we consider the R—H...B interaction where the three electrons are taken into account, each having spin function α or β, than for the α-type hydrogen electron we have β electrons of R and B. In the same way for β electron of hydrogen atom there are the same type α-electrons for R and B. Hence for both cases the R...B interaction is realized by a triplet state.

The BEBO method has been developed for a lot of chemical problems, for example in connection with the Intersecting-State Model[42] to estimate activation barriers and reactive bond distances for reactions concerning the transfer of hydrogen atoms.

It has been pointed out that the ideas of BEBO method may be the way for the description of O—H...O bonds.[43] Combining Johnston and Parr's empirical equation (Eq. 35) with Pauling's bond number definition (Eqs. 1 or 4) the equation was obtained which enables the H...O interaction to be estimated from H...O interatomic distance:

$$E(n) = E(1)\exp[\alpha(r_S - r_n)] \tag{39}$$

r_S is, as for the previous equations, the single bond length for which the n-bond number equals the unity and its energy amounts to $E(1)$. $E(n)$ is the energy of the bond of r_n length. The constant value α equals to p/c and for the H...O interaction $\alpha = 4.338$ Å^{-1}. The α-value may be calculated from Eq. 39 if we know $E(1)$, the length of the single bond—r_S and additionally the bond energy for the bond of the known length. For H...O distance of 1.22 Å corresponding to the bond of the bond number equal to 0.5 we have the energy of 150 kJ/mol. Than for H...O interaction Eq. 39 becomes:[43]

$$E(n) = -468.9\exp[4.338(0.957 - r_{H...O})] \tag{40}$$

In order to check up validity of this approach a comparison with the results obtained with the use of Lippincott–Schroeder potential is presented. The Lippincott and Schroeder term describing H...O interaction has the form:

$$E_{LS} = -D^*\exp\left[\frac{-n^*(r_{H...O} - r_0^*)^2}{2(r_{H...O} - r_0^*)}\right] \tag{41}$$

D^*, r_0^* and n^* are the constants describing the H...O contact and related to the constants D, r_0, n describing the O—H bond within O—H...O system.

To compare the Lippincott and Schroeder model of hydrogen bonding with the approach of Johnston and Parr,[14] O—H...O systems were taken

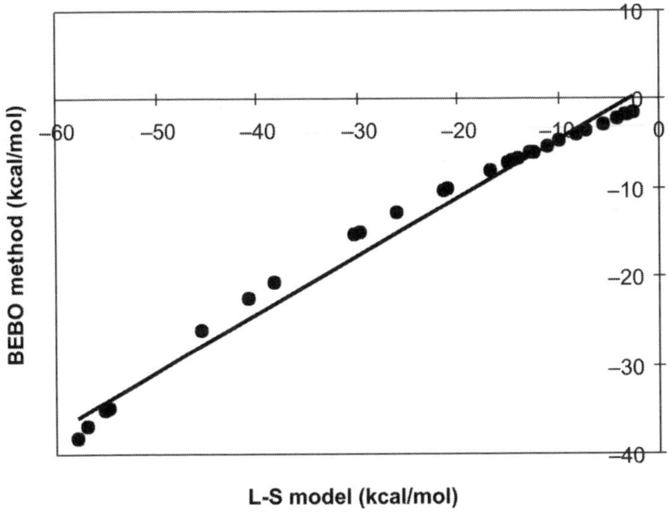

Figure 7. The correlation between Lippincott and Schroeder energies of H...O contacts and the energies obtained from BEBO method within the model.

from Cambridge Structural Database (CSD).[44,45] Only the data of the accurate ($R \leq 5\%$, e.s.d's$\leq$0.003 Å) neutron diffraction structures were taken into account. To avoid the effect of the inhomogenity of the sample, only the C=O...H—O—C systems are considered. The set of 30 O—H...O intermolecular H-bonds was obtained. Figure 7 presents the relationship between H...O energies of that sample, obtained from Lippincott and Schroeder model (Eq. 41) and the energies of the same systems obtained from Eq. 40.

The linear regression for the data of Figure 7 (presented at figure) is:

$$E(n) = 0.656E_{\mathrm{LS}} + 2.108\,(\text{in kcal/mol}) \tag{42}$$

with the linear correlation coefficient $r = 0.993$. The similar results were obtained earlier[43] for another data set taken from CSD. It is clear from a visual observation that the line in Figure 7 is rather curved. Thus for the sample presented in Figure 7 the polynomial regression line was also constructed:

$$E(n) = -0.0057E_{\mathrm{LS}}^2 + 0.3057E_{\mathrm{LS}} - 0.9698\,(\text{in kcal/mol}) \tag{43}$$

and the correlation coefficient $r = 0.999$ was calculated.

It may be concluded that the much simpler formula (Eq. 40) may replace a more complex Lippincott–Schroeder potential in most qualitative discussions based on X-ray or neutron difraction crystal structure and molecular geometry. The study of the effect of through-resonance on the geometry of *p*-nitrosophenolate anions[46] may be an example. Since it was pointed out that in the description of structures and properties of *p*-substituted benzenes an important contribution arises from the canonical structure with a full charge transfer from an electron donating to an electron accepting group,[47] the contribution of that canonical quinoid structure was investigated for some benzene derivatives. Hiberty and Ohanessian[48] using VB calculations showed that quinoid structure is negligible for *p*-nitroaniline and *p*-nitrophenol (less than 1% contribution), while the study of X-ray crystal structure of *N*,*N*-dialkyl-*p*-nitroaniline[49] showed the contribution of that structure is more significant.

For the crystal structures of sodium *p*-nitrosophenolate trihydrate[50] and magnesium *p*-nitrosophenolate hexahydrate[51] the BEBO model was applied to explain the geometrical parameters of molecules.[46] For both crystals the contribution of the quinoid structure is significant, for sodium and magnesium salts it is 35.9% and 22.5% respectively. It seems that the greater contribution of the quinoid structure, i.e. with charge transfer from an electron donating substituent (O^-) to an electron accepting substituent (nitroso group), for sodium salt is reflected by the intermolecular interactions, particularly those which are the strongest—H-bonds. We observe the greater H-bond interactions for sodium salt because the Lewis basic character for nitroso substituent is greater here than for nitroso group within magnesium derivative structure. The greater basic character reflects the greater charge transfer from O^- to NO group. In the case of sodium salt, the nitroso substituent is involved in four H-bond interactions, the total interaction energy equals to 5.1 kcal/mol, whereas there is only one H-bond for O^- equal to 1.0 kcal/mol. In the case of magnesium derivative there are three H-bonds for NO-group of the energy of 3.8 kcal/mol and three H-bonds for O^- of 4.4 kcal/mol. The H-bond energies for both groups of magnesium salt are of similar magnitude what reflects in the smaller charge transfer from O^- to the NO group and in consequence in the lesser contribution of the quinoid structure than for the sodium salt. The H-bond energies for these structures[46] were calculated using Eq. 40.

VII. THE VALENCE SUM RULE AND ITS APPLICATIONS TO WEAK HYDROGEN BONDINGS

The bond valence model was applied for the investigations of the relationships between geometrical parameters in strong and medium hydrogen bonds, for example O—H...O and N—H...N systems.[2,8,12,16,18,29,43,52] The relationships between X—H covalent bond length and H...Y distance for O—H...O and N—H...N bonds were shown in the previous section. The appropriate dependences obtained from the valence sum rule are in agreement with the neutron diffraction data showing that BV model allows to predict the elongation of covalent bond for hydrogen bonded systems.

The aim of this section is to show if the dependences obtained from the BV model for weak H-bondings are in line with the other results.

It was pointed out[53] that there is a correlation between the covalent bond length of C—H and the H...O distance for C—H...O systems. However the studied correlation is not linear and is not of the form of a simple function: exponential or quadratic. Additionally the sample was chosen to avoid the effects of chemical inhomogenity, thus the C—H...O systems of α-amino-acids structures were taken into account.[53] However it is very interesting to know if the relation C—H bond length versus H...O distance obtained from the valence sum rule may be useful to predict the elongation of the length of C—H covalent bond for any kind of C—H...O system.

Figure 8 shows a plot of C—H bond length against hydrogen bond distance H...O for the data sample containing 96 C—H...O systems taken from CSD ($R \leq 5\%$, e.s.d's$\leq$0.005 Å). The sample contains bonds with C(sp^2)—H donors and for H...O distances not greater than 2.8 Å. The distance criterium was chosen to take into account only stronger C—H...O

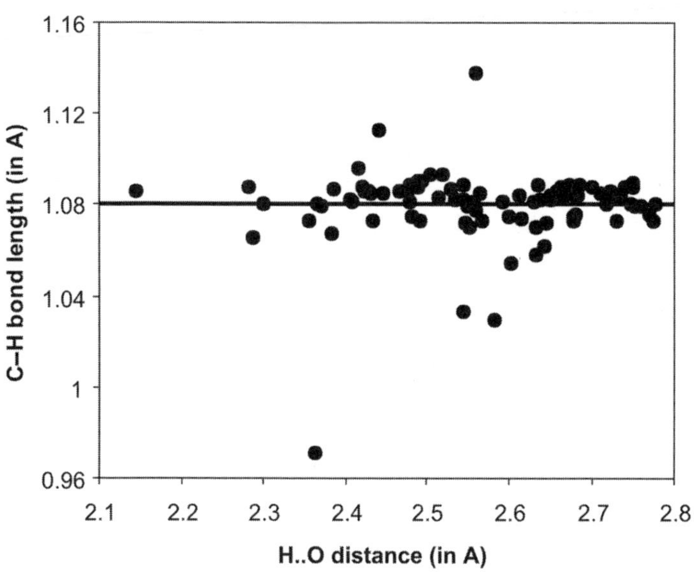

Figure 8. Scatter plot of the length of C—H against H...O distance within C(sp^2)—H...O systems (taken from ref. 12).

bonds. Additionally it is well known[54] that the strengths of C—H...O bonds increase for C—H donors in the following order: C(sp^3)—H, C(sp^2)—H, C(sp)—H.

There are no appropriate neutron diffraction results on C(sp)—H...O systems in Cambridge Structural Database, so C(sp^2)—H...O species were taken into account.[12] It is well known that the benzene C(sp^2)—H bond often acts as a proton donor[55,56] therefore only this kind of systems and related ones with C(sp^2)—H are presented here.

Solid line in Figure 8 represents the bond length of 1.08 Å. This value was proposed by Jeffrey and Lewis[13] for such cases where the proton position within C—H...Y bond is not accurately determined. It is a common procedure for the X-ray diffraction results. Jeffrey and Lewis proposed[13] the evaluation of H-atom position for X-ray hydrogen bonding, H-atom is moved along X—H bond to the standard length (O—H–0.98 Å; N—H–1.01 Å; C—H–1.08 Å). Figure 8 shows that the neutron diffraction results on C—H...O bonds are dispersed near the solid line so there is no reason to apply the valence sum rule for C(sp^2)—H...O systems. The same may be expected for weaker C(sp^3)—H...O bonds. The case of C(sp^2)—H...O systems show that for the weak H-bonds the dependences obtained from BV model are not in agreement with the accurate neutron diffraction results. It may be caused by the intermolecular interactions in crystals which strongly influence the geometry of H-bonding and which are more important for weaker bonds like C—H...C, C—H...O than for medium and strong N—H...N and O—H...O interactions.

The valence sum rule was also applied for unconventional interactions—dihydrogen bonded (DHB) systems. Such interactions are mainly present in transition-metal compounds[57–59] and are designated as M—H...H—A, where M is a metal and H—A represents the conventional donating bond as O—H and N—H. M—H may be treated as a proton accepting group as for a lot of transition metal compounds where there is a negative net charge on hydrogen atom —M—$H^{-\delta}$.

DHB bonds were also investigated using ab initio methods. For example, simple systems like Li—H...H—F, BeH_2...H—F and the other similar complexes were investigated theoretically[60–63] but also the complexes known from the experimental measurements were optimized.[60,64–67]

Figure 9 shows the dependence between H...H distance and an elongation of H—F bond for A—H...H—F systems. H—F molecule was chosen as a rather strong donating system, A—H designates an accepting bond.

Solid line in Figure 9 was obtained from the valence sum rule applied for hydrogen atom:

$$s_{H...H} + s_{H\text{–}F} = 1 \tag{44}$$

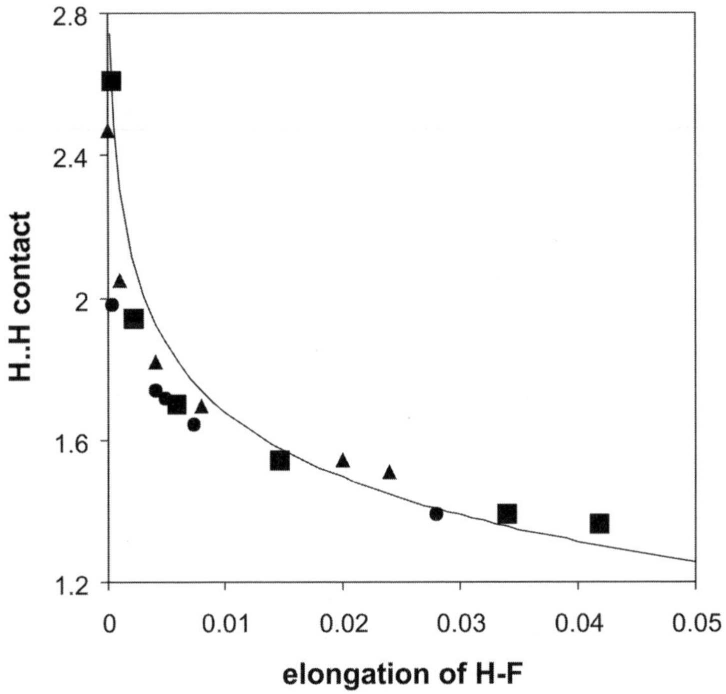

Figure 9. The relative elongation of H—F bond (in A) versus the H ... H distance (in Å) for DHB systems (taken from ref. 63).

The constant values appropriate for the calculation of bond valences were given by Dunitz[11] and they are presented in Table 1. Squares and triangles in Figure 9 correspond to A—H ... H—F dihydrogen bonded complexes optimized within MP2/6–311++G** and MP2/6-31G** levels of theory respectively. For these two levels of theory the same series of complexes; H—F donating molecule and CH_4, SiH_4, BeH_2, MgH_2, NaH, LiH accepting molecules within complexes were investigated. Circles correspond to the series optimized within B3LYP/Lanl2DZ level, where H—F is again chosen as donating molecule for DHB complexes and as accepting molecules CH_4, SiH_4, GeH_4 and SnH_4 were taken into account as the homogenous species corresponding to the fourth main group of the periodic table of elements. Additionally for this level of theory the CrH_2 ... HF complex was optimized as one corresponding to transition-metal complexes detected as DHB systems by spectroscopic and neutron diffraction techniques.[57–59]

Figure 9 shows that ab initio and DFT theoretical results are in good agreement with the empirical curve (represented by the solid line) obtained from Eq. 44—the valence sum rule. The difference between DHBs complexes and C—H...O and the other weak hydrogen bonds in connection with BV model may be explained in the following way;

- for DHBs the homogenous series was considered (the same H—F donating molecule and the similar accepting molecules);
- there is no other intermolecular interactions for DHB complexes, opposite to the neutron diffraction results for C—H...O bonds, where the intermolecular interactions in the crystal lattice may strongly influence on C—H...O geometry.

VIII. THE STUDY OF REACTION PATHS FROM THE CRYSTAL STRUCTURE DATA

It has been pointed out that the various crystal structures of homogenous compounds showing the same, common molecular fragment may be treated as points yielding the reaction path.[68] Thus some of correlations between geometrical parameters of molecules taken from the neutron diffraction and X-ray crystallographic data may reflect the changes in the molecular structure during the chemical reactions. The correlations between molecular parameters may be predicted by the bond valence model, particularly by the valence sum rule—Eq. 6 (or its appropriate form for the triatomic systems—the bond number conservation rule—Eq. 3). The reaction paths as determined from the crystal structure data were investigated very early by Bürgi and Dunitz[10,11]—nucleophilic addition to carbonyl groups, nucleophilic substitution at tetrahedral molecules, electrocyclic reactions, intramolecular processes with small energy barriers and with low activation energies were taken into account.

Since there are many review articles concerning the transformation paths for different types of chemical reactions: organic addition and elimination reactions of carbonyl derivatives,[69] ligand rearrangement and substitution reactions of transition metal complexes,[70] nucleophilic substitution (S_N2) reactions,[71] these problems are not subject of this article. We only mention an example to give the Reader a chance of understanding the model and its application. In the case of nucleophilic substitution within $XCdS_3Y$-fragments the appropriate reaction path may be modeled according to the data taken from CSD:[44,45]

$$Y + S_3CdX \rightarrow Y \ldots S_3Cd \ldots X \rightarrow YCdS_3 + X$$

It is an organic S_N2 reaction: attack of a nucleophile Y opposite to a group X, formation of a new bond between Y and cadmium molecule, breaking of the bond to X and inversion of configuration at the reaction center. The following equation based on the bond number conservation rule allows to create the corresponding reaction path:

$$10^{-\Delta X/a} + 10^{-\Delta Y/a} = 1 \tag{45}$$

where ΔX and ΔY are the differences between the observed distances Cd...X and Cd...Y and appropriate reference distances. For the $XCdS_3Y$ fragment constant a equals to 1.05 Å. The good agreement between BNCR (Eq. 45) and experimental data was obtained.[71] Figure 10 shows the dependence between ΔX and ΔY obtained from Eq. 45 for $XCdS_3Y$ fragment (solid line). The same relation for $XSnC_3Y$ is present (broken line), ΔX and ΔY correspond to differences between X...Sn, Y...Sn and appropriate reference distances here. The constant a for $XSnC_3Y$ fragment equals to 1.20 Å.

As it is well known[10] the correlations between structural parameters of molecules are reminiscent of the structural changes occurring during chemical reactions. The approach proposed by Bürgi and Dunitz[10,11] may be applied to many different reactions. For example such treatment was applied to the study of the reaction path of proton transfer within C=O...H—O system.[72] The following reaction corresponds to this

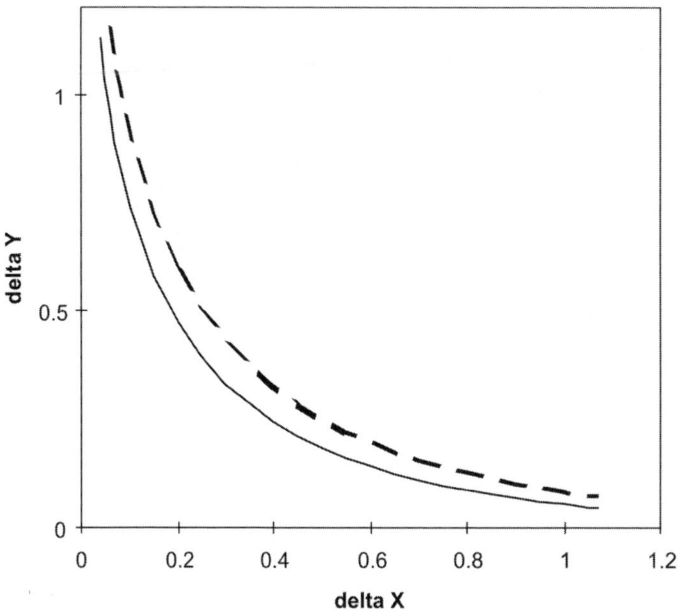

Figure 10. The dependences between ΔX and ΔY for $XCdS_3Y$ (solid line) and $XSnC_3Y$ (broken line) fragments obtained from BNCR.

fragment:

$$—C{=}O\ldots H{—}O{—}C{\rightarrow}C{—}O{—}H\ldots O{=}C$$

The neutron diffraction (ND) data of the high accuracy was taken from CSD (e.s.'s≤0.005 Å and R≤8%). The use of ND results but not of X-ray measurements was very important here since it is well known that X-rays are scattered at the electron shells and the positions of H-atoms are not well determined. Neutrons are scattered at the atomic nuclei and hence the positions of H-atoms are of similar accuracies as the other atoms.

Figure 11 shows the dependence between the positions of protons measured from the middle of O...O distance, delta *r*, and the O...O distance, they are denoted by full circles. Solid line was obtained from BNC rule (Eq. 3) and is in a good agreement with ND results, particularly for the shorter O...O distances.

The solid line was obtained in accordance with the constants of Dunitz[11]—the length of the single O—H bond r_0=0.957 Å and $r_{1/2}$=1.22 Å whereas the broken line corresponding to the constants proposed by Gilli

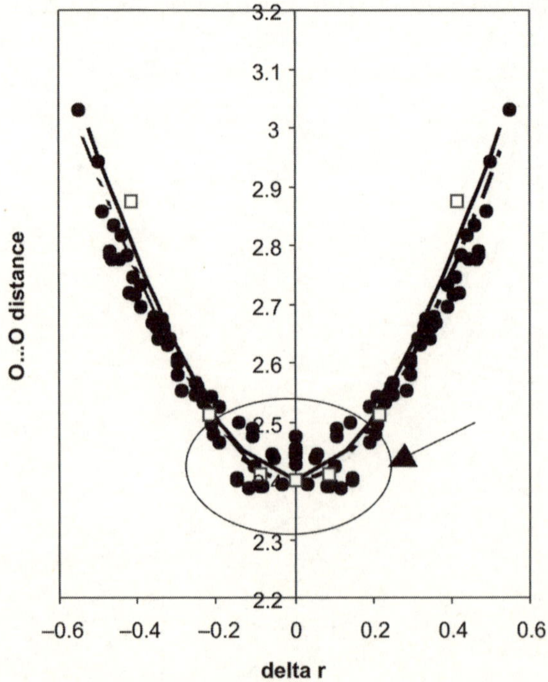

Figure 11. The reaction path for the proton transfer within C=O...H—O systems, based on data from ref. 72.

and coworkers[8]: $r_0 = 0.925$ Å and $r_{1/2} = 1.22$ Å. Gilli and coworkers claimed that such approach is in a better agreement with the character of interactions within O—H...O systems. It means that the electrostatic part of interaction is greater for the longer O...O distances and the use of the other constants takes such character into account. The broken line in Figure 11 confirms that because it is in a better agreement with ND data than the solid line.

Figure 11 may be understood in the following way. The contour of the experimental neutron diffraction results represents possible geometries of O—H...O bonds during the proton transfer. The contour is in line with the high level ab initio results on formic acid dimer.[73] Empty squares in Figure 11 represent the pre-tunneling, post-tunneling and energetic stable configurations of formic acid dimer (FAD). The same or similar configurations during the proton transfer exist for the other carboxylic acids, for example, the benzoic acid dimer in the solid state was investigated extensively.[74]

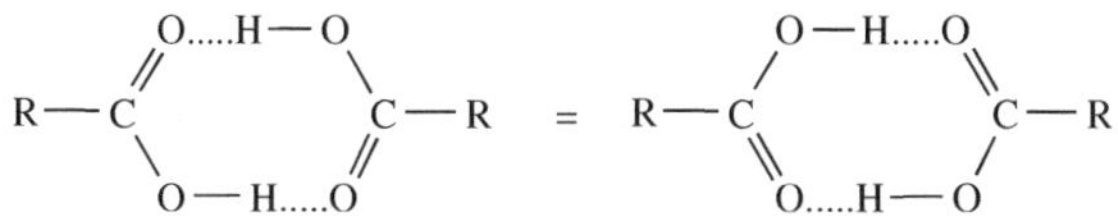

It was demonstrated[72] that the proton transfer reaction path presented in Figure 11 may also approximately show the potential barrier height for this process. The greater are the differences between the configurations during the proton transfer, it means the greater the way within the contour in Figure 11 the greater is the barrier height. Hence the distribution of theoretical points representing FAD configurations (Figure 11) confirms the high barrier of 8.9 kcal/mol as it was calculated by Kim.[73] Figure 11 also presents the region of the so-called low-barrier hydrogen bonds (LBHB) for which the potential energy barriers are of about ~1 kcal/mol or even less, in the extremal case there is no barrier.[75,76] For O...O distances shorter than 2.5 Å there is a continuum of neutron diffraction points (full circles) in the LBHB region indicated in Figure 11 by the arrow.

The similar reaction path obtained from ND results was created for O—H...O bonds and for deuterated systems—O—D...O bonds. It was pointed out[32] that the comparison of O—H...O geometries with O—D...O ones is in line with the geometrical changes known as Ubbelhode effect.[77,78] Additionally the relationships between geometrical parameters may be obtained for O—H...O bonds as well as for O—D...O systems and both these relations are in agreement with the bond number conservation rule.[32]

A similar reaction path has been also studied for N—N...N[79] and N—H...O (N...H—O)[80] bonds. However for these cases the correlations between the coordinates q_1 and q_2 were investigated:

$$q_1 = \frac{1}{2}(r_1 - r_2) \quad \text{and} \quad q_2 = r_1 + r_2 \tag{46}$$

where r_1 is the length of N—H or O—H covalent bond and r_2 is H...N or H...O distance. For linear hydrogen bonds q_1 corresponds to Δr used for the study of C=O...H—O system (Figure 11) and q_2 to R-distance (O...O distance in Figure 11). For the sets of results containing not only linear systems the use of q_1 and q_2 coordinates seem to be more proper than the use of Δr and R. For the investigation of the proton transfer for C=O...H—O fragments only linear and approximately linear systems were taken into account[72] thus the use of Δr and R coordinates was justified.

The proton transfer reaction paths for N—H...N and N—H...O (N...H—O) bonds were investigated using q_1 and q_2 coordinates. For N—H...N bonds the solid state ^{15}N NMR results, MP2/6–31+G(d,p) calculations and the results from bond number conservation rule were compared showing the agreement between them.[79] The agreement between BNCR and experimental data was also achieved for N—H...O bonds.[80]

IX. THE BOND-VALENCE MODEL AND LEWIS ACIDITY/ BASICITY STRENGTHS

It has been pointed out that the stable complexes are formed only between compounds where Lewis acid strength and Lewis base strength are approximately equal.[2,81] This theorem is in line with the Donohue rule[82] which was extended and applied in a wider topological context by Etter[83] that H-bonds are formed between "good proton donors" and "good proton acceptors". Etter claimed that strong proton donors such as those in carboxylic acids, imides, phenols, amides etc. are most often involved in H-bonds in crystals. If all more acidic protons form H-bonds and the extra proton acceptors are available then the less acidic groups like C—H of acetylenes and aromatic compounds may form H-bondings. Hence C—H...O bonds are often detected in crystals.[84]

The theorem of Lewis acidity and basicity named as the valence-matching principle[2] comes from the bond valence model since the valence of a bond formed by a cation corresponds to Lewis acid strength and the Lewis base strength corresponds to the valence of a bond formed by an anion. It was shown[2] that the average value $\langle s \rangle$ of the bonds formed by a cation may be expressed:

$$\langle s \rangle = \frac{V}{\langle n \rangle} \tag{47}$$

where V is the atomic valence (oxidation state) and $\langle n \rangle$ is the average coordination number of a cation found in a large number of stable solids. $\langle s \rangle$ is close to the valence of each of the bonds formed by the cation and it correlates with electronegativity[85] thus it may be treated as Lewis acid strength—s_a.

$$s_a = \langle s \rangle = \frac{V}{\langle n \rangle}, V > 0 \tag{48}$$

V is positive as the oxidation state of the cation. A similar approach may be used to define a Lewis base strength—s_b.

$$s_b = \langle s \rangle = \frac{V}{\langle n \rangle}, V \leq 0 \tag{49}$$

V is negative because the oxidation state of an anion is negative. Besides anions form bonds with a wider range of coordination numbers and hence with a wider range of valences than cations. Brown[2] defined the so-called maximum base strength:

$$s_m = \frac{V}{n_{min}} \tag{50}$$

n_{min} is the smallest possible coordination number. Tables 4 and 5 present some Lewis acid and base strengths.

The approach presented here was applied for the study of Lewis basicity of the acetate and trifluoroacetate ions.[81] The formal charge of both ions is -1 thus the following equation may be written:

$$\Sigma s_b - \Sigma s_a = 1 \tag{51}$$

From the crystal structures of acetates and trifluoroacetates the geometry of COO^- group is known and hence it is possible to estimate that trifluoroacetate ion is a weaker Lewis base than the acetate ion.[81] The base strength of the two O atoms is 1.1 valence units for the acetates and 0.9 valence units for trifluoroacetates. Such situation follows directly from Eq. 51 since there are basic —CF_3 groups for trifluoroacetates and weak acidic —CH_3 groups for acetates.

The basicities of acetates and trifluoroacetates were determined from the geometry of their molecules known from X-ray diffraction measurements. It is important that the structural information may be a source of acid and

Table 4. Lewis acid strengths in v.u. (valence units)—taken from ref. 86

Li	Be	B	C	N	O	F
0.205	0.501	0.87	1.35	1.67		
Na	Mg	Al	Si	P	S	Cl
0.156	0.334	0.57	1.00	1.25	1.5	1.75
K.	Ca	Ga	Ge	As	Se	Br
0.126	0.274	0.65	0.89	1.13	1.5	
Rb	Sr	In	Sn	Sb	Te	I
0.120	0.233	0.50	0.68	0.83	1.20	1.2
Cs	Ba	Tl	Pb	Bi^{3+}		
0.113	0.195	0.49	0.70	0.48		
Sc^{3+}	Ti^{4+}	V^{5+}	Cr^{6+}	Mn^{3+}	Fe^{3+}	Co^{3+}
0.49	0.67	1.08	1.50	0.52	0.53	0.51
	Mn^{2+}	Fe^{2+}	Co^{2+}	Ni^{2+}	Cu^{2+}	Zn^{2+}
	0.34	0.34	0.35	0.34	0.39	0.40

Table 5. Lewis base strengths in v.u. (valence units)—taken from ref. 2

s_b	s_m	*Anion*
0.50	2.00	O^{2-}
0.33	1.00	SiO_4^{4-}, BO_3^{3-}
0.25	0.75	PO_4^{3-}
0.22	0.67	CO_3^{2-}, HPO_4^{2-}
0.21	1.00	F^-
0.17	0.50	SO_4^{2-}, Cl^-, $H_2PO_4^-$
0.17	0.34	H_2O
0.11	0.33	NO_3^-
0.09	0.50	Br^-
0.08	0.25	ClO_4^-, BF_4^-
0.06	0.50	I^-

base strength. It was found that the base strength, s_b is related to the acid pK value by the relationship:[81]

$$s_b = 0.42 + 0.032 \text{ pK} \tag{52}$$

For s_b equal to 0.55 for the acetate and $s_b = 0.45$ for the trifluoroacetate there are 4.1 and 0.9 values of pK respectively,[81] while the experimental values are 4.8 and 0.3.

The analysis of the Lewis basicity of C═O group and Lewis acidity of O—H bond has been performed for —C═O...H—O— systems existing in organic crystal structures.[16] According to Donohue's theorem[82] and Etter's rule[83] the base strength of C═O should correspond to O—H acid strength for these systems. Figure 12 presents such dependence. As one can see there is no strong correlation between basicity of C═O and acidity of O—H, the correlation coefficient for this dependence is 0.692. However the results show that acidity and basicity values are approximately equal.

s_a acidity and s_b basicity were defined here in a different way than the proposals of Brown (Eqs. 48–50). The Lewis acidity s_a of O—H bond equals to $(1 - s_{OH})$. s_{OH} is the bond valence and its decrease is connected with the ability to form intermolecular connections in crystals. Such definition directly originates from the valence sum rule—the atomic valence of hydrogen equals unity and the elongation of O—H bond causes that

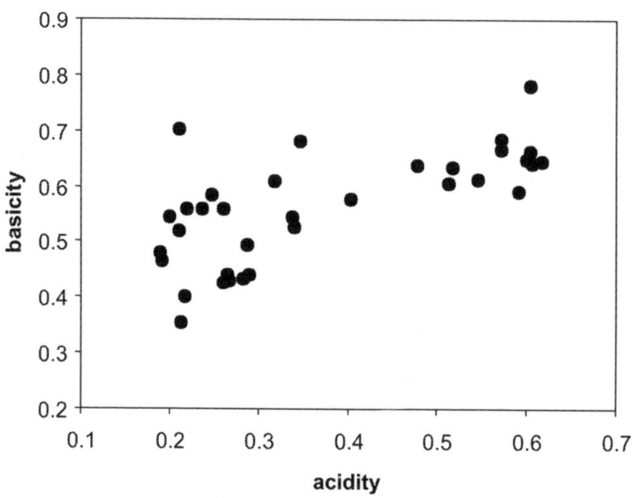

Figure 12. The Lewis acidity of O—H bond versus the Lewis basicity of C═O carbonyl group for 47—C═O...H—O— systems taken from CSD (the picture taken from ref. 16).

external bonds (intermolecular contacts) may be formed and therefore the valence sum rule is in force.

The oxidation state of oxygen atom equals to −2, hence the Lewis basicity of the carbonyl group s_b equals to $(2 - s_{C=O})$, $s_{C=O}$ is the bond valence of C=O bond.

It was shown in the previous sections how to calculate s_{OH}-valences. We may write for the carbonyl group:

$$s_{C=O} = \exp\left[\frac{(r^0_{C-O} - r_{C=O})}{B_{C-O}}\right] \tag{53}$$

where $r^0_{C-O} = 1.428$ Å is the reference length of the single C—O bond (it is the electron diffraction value for methanol), B_{C-O} is the constant value calculated from the equation:

$$2 = \exp\left[\frac{(1.428 - 1.209)}{B_{C-O}}\right] \tag{54}$$

where $r_{C=O} = 1.209$ Å is the bond length for pure double C=O bond (data of acrolein—ref. 87).

In the previous sections of this review the application of the valence sum rule (VSR) for triatomic systems–hydrogen bonds was shown, particularly for H-atom within H-bonds. VSR may be also applied for oxygen atom of C=O...H—O system:

$$s_{H...O} + s_{C=O} = 2 \tag{55}$$

Hence the dependence between H...O distance and C=O bond length may be derived (Figure 13).

Figure 13 also shows the points corresponding to neutron diffraction results of high precision ($R \leq 8\%$; e.s.d's ≤ 0.005 Å) taken from Cambridge Structural Database. Circles in Figure 13 represent O—H...O=C systems, triangles $COO^-...H^+...{}^-OOC$ ones, the solid line shows the curve obtained from Eq. 55. The agreement between the valence sum rule and neutron diffraction results is satisfactory.

X. THE BOND NUMBER CONSERVATION RULE AS A TOOL FOR THE DETERMINATION OF STABLE H-BONDED SYSTEMS

There are different semiempirical models which allow to estimate the H-bond energy. The most known and the most often used one have been proposed by Lippincott and Schroeder.[30,31] This model was applied for

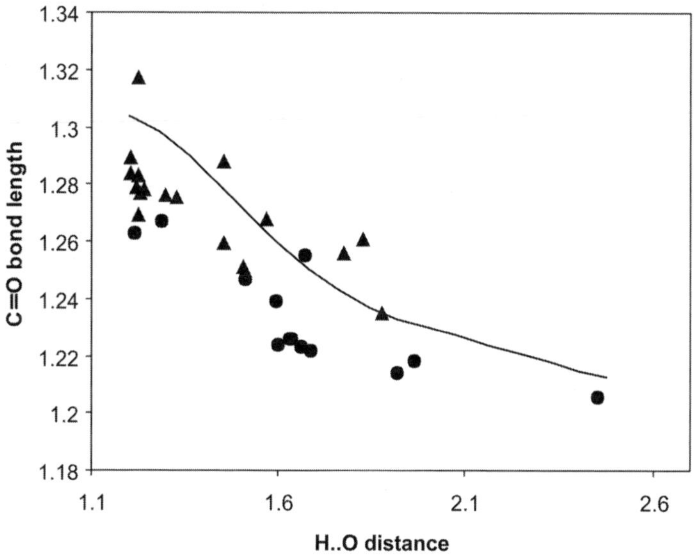

Figure 13. The relationship between H...O distance (in Å) and C=O bond length (in Å) (the picture taken from ref. 16).

different types of H-bridges, like O—H...O, N—H...N, N—H...O etc., but most often it is treated as a useful tool for the description of O—H...O systems. For this model, for O—H...O bond, one can see that the H-bond energy E_{HB} is the sum of three terms:

$$E_{HB} = E_{OH} + E_{H...O} + E_{O...O} \tag{56}$$

where E_{OH} is the term connected with O—H covalent bond, $E_{H...O}$ describes H...O interaction and $E_{O...O}$ is for the interaction between terminal, for O—H...O bridge, oxygen atoms. The terms were originally defined by Lippincott and Schroeder[30,31] in the following way:

$$E_{OH} = D\left[1 - \exp\left(\frac{-n\Delta r^2}{2r}\right)\right] \qquad \Delta r = r - r_0 \tag{57}$$

$$E_{H...O} = -D^*\left\{\exp\left[\frac{-n^*\left(R - r - r_0{}^*\right)^2}{2(R - r)}\right]\right\} \tag{58}$$

$$E_{O...O} = A\exp(-bR) - \frac{B}{R^m} \tag{59}$$

where: n, n^*, D, D^*, A, B, b, m are constants, r_0 is the length of reference single O—H bond, r is the length of O—H bond for the considered

O—H...O system and R is the O...O distance for this system. The following constant values are often used in considerations on O—H...O bonds: D is the dissociation energy of O—H bond and it is equal to 118.0 kcal/mol, n=9.18, A=43.00 × 10^5, b=4.815, r_0=0.957 Å, B=18.2; $D^*=D/g$ and $n^*=g\,n$, where g=1.45.

Such approach does not allow to calculate correctly energies of H-bonds for which O...O distances are short. For short O...O distances, in the region of 2.4–2.55, the O—H...O systems are not stable within the original Lippincott–Schroeder model. However it is known from different investigations that such systems known as low barrier H-bonds (LBHB) are stable and their H-bond energies are of about 30 kcal/mol.[75,76] Gilli and coworkers[8] have pointed out that some of the constants of the Lippincott–Schroeder model should be changed to correct H-bond energies for LBHBs. It is particularly connected[8] with the equalization of O—H bond and of H...O distances for short O...O distance what is the effect of covalency of H...O interaction. The appropriate corrections for g and B constants have been proposed.[52]

$$\begin{aligned} g &= 1 + 1.8(R - 2.40) \quad \text{for} \quad R \in \langle 2.40; 2.65 \rangle \\ B &= 72.8(R - 2.40) \quad \text{for} \quad R \in \langle 2.40; 2.65 \rangle \end{aligned} \tag{60}$$

For R> 2.65 Å B equals to 18.2 and g to 1.45 while for R< 2.40 Å B is equal to zero and g to 1. In this way the modified model of Lippincott and Schroeder (MLS) has been introduced to correct H-bond energies for short O...O distances.[52]

The MLS model in connection with the bond number conservation rule (BNCR) may be used for the estimation of the stability of H-bonds. It is known that the sum of bond numbers of O—H bond and H...O contact should be equal to unity (Eq. 21). The systems for which this sum is far away from unity are not stable, hence the H-bond energies for them should have the great positive values.

The following treatment was proposed[52] to consider O—H...O bonds. If we know the O...O distance for O—H...O system, thus it is possible to predict the position of H-atom using the BNCR. It is the useful tool for the X-ray data for which the positions of H-atoms are not accurate. However for the fixed O...O distance we may have different H-atom positions fulfilling the BNCR. But the calculation of H-bond energies within MLS allows to take out the system fulfilling BNCR and of the lowest energy. Figure 14 shows the relationship between such obtained energies and corresponding O...O distances (continuous line). The solid line of Figure 14 represents the O—H...O bridges for which BNCR is in force.

Additionally the points (circles) representing neutron diffraction structures of the high precision are shown in Figure 14. The experimental data

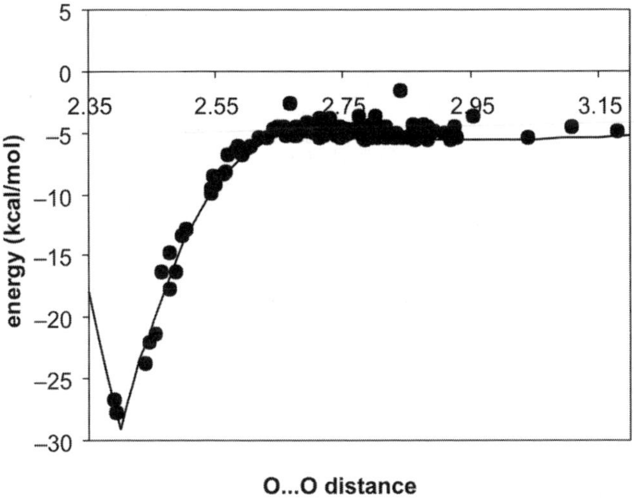

Figure 14. The relationship between O...O distance (in Å) and H-bond energy (taken from ref. 52).

are in line with the solid line, it means that for the crystal structures, O—H...O systems try to have the geometries which are in agreement with the bond number conservation rule and which correspond to the lowest energies. This finding is of the great importance for investigators considering H-bonds because the main source of H-bond geometries is CSD for which over 99.5% of all data become from X-ray diffraction measurements and less than 0.5% from neutron diffraction results. Thus for a lot of compounds there are no results with accurate H-atom positions. The approach presented above allows to estimate H-bond energy knowing only O...O distance—the polynomial function may be fitted to the solid line of Figure 14.[52]

XI. CONCLUSIONS

All chemical and most of the physicochemical properties of molecules may be predicted provided the wave function is known. Unfortunately for many chemically and biochemically interesting systems high level ab initio or DFT calculations can not be carried out for the complexity of the systems in question. Here the role of empirical models is very important. They may be tested either by comparison with experimental data or with high level calculated results. If the empirical models have acceptable agreement with the other data, they may be a very useful tool for studying the relations between structure and chemical/physicochemical/biochemical properties of molecular systems in question.

The concept of bond number/bond valence and related ones belongs to the class of "primitive" pattern of understanding,[88] the term introduced by C. A. Coulson for the concepts which are everyday tools of chemists in their attempts to understand the chemical and related interactions.

REFERENCES

1. Pauling, L. *J. Am. Chem. Soc.* **1947**, *69*, 542.
2. Brown, I. D. *Acta Cryst.* **1992**, *B48*, 553–572 and references cited therein; Brown, I. D. Bond-Length–Bond-Valence Relationships in Inorganic Solids, In *Structure Correlation*, Bürgi, H.-B.; Dunitz, J. D. Eds; Weinheim, **1994**.
3. Wells, A. F. *Structural Inorganic Chemistry*, Oxford University Press, **1984**; Cotton, F. A.; Wilkinson, G. *Advanced Inorganic Chemistry*, Interscience, **1972**.
4. There are numerous definitions of bond order in the frame of π-electron models, see: Cioslowski, J.; Mixon, S. T. *J. Am. Chem. Soc.* **1991**, *113*, 4142–4145 and references cited therein, for example Wiberg's model—Wiberg, K. *Tetrahedron*, **1968**, *24*, 1083, see also textbooks on advanced quantum chemistry (for example; McWeeny, R. *Coulson's Valence*, Oxford University Press, **1979**).
5. Jenkins, H. O. *J. Am. Chem. Soc.* **1955**, *77*, 3168.
6. Johnston, H. S. *Adv. Chem. Phys.* **1960**, *3*, 131.
7. Speakman, J. C. *Molecular Structure by Diffraction Methods*, Sim, G. A.; Sutton, L. E. Eds.; *3*, The Chemical Society, London, **1975**.
8. Gilli, P; Bertolasi, V.; Ferretti, V.; Gilli, G. *J. Am. Chem. Soc.* **1994**, *116*, 909–915.
9. Brown, I. D.; Altermatt, D. *Acta Cryst.* **1985**, *B41*, 244–247.
10. Bürgi, H.-B. *Angew. Chem. Internat. Edit.* **1975**, *14*, 460.
11. Dunitz, J. D. *X-Ray Analysis and the Structure of Organic Molecules*, Cornell University Press, Ithaca, **1979**.
12. Grabowski, S. J. Simulations of Hydrogen Bonds in Crystals and Their Comparison with Neutron Diffraction Results, In *Neutrons and Numerical Methods—N_2M*, Grenoble, France, December **1998**, AIP Conference Proceedings 479, Johnson, M. R.; Kearley G. J.; Büttner, H. G. Eds.; American Institute of Physics **1999**; Grabowski, S. J.; Krygowski, T. M.; Stepien, B. *J. Phys. Org. Chem.* **2000**, *13*, 740.
13. Jeffrey, G. A.; Lewis, L. *Carbohydr. Res.* **1978**, *60*, 179.
14. Johnston, H. S.; Parr, Ch. *J. Am. Chem. Soc.* **1963**, *85*, 2544–2551.
15. Gilli, G.; Bertolasi, V.; Veronese, V. *Acta Cryst.* **1983**, *B38*, 450.
16. Grabowski, S. J. *Tetrahedron* **1998**, *54*, 10153–10160.
17. Brown, I. D. *Chem. Soc. Rev.* **1978**, *7*, 359.
18. Steiner, T. *J Phys. Chem. A* **1998**, *102*, 7041.
19. Bergerhoff, G.; Hundt, R.; Sievers, R.; Brown, I. D. *J. Chem. Inf. Comput. Sci.* **1983**, *23*, 66.
20. Bergerhoff, G.; Sievers, R. *Nachr. Dokum.* **1989**, *40*, 27.
21. Vajda, E.; Hargittai, I. *J. Phys. Chem.* **1992**, *96*, 5843. Partalone, G.; Schultz, G.; Domenicano, A.; Hargittai, I. *Chem. Phys. Lett.* **1992**, *197*, 482.
22. Legon, A. C.; Millen, D. J. *Chem. Phys. Lett.* **1988**, *147*, 484.
23. Krygowski, T. M.; Grabowski, S. J.; Konarski, J. *Tetrahedron* **1998**, *54*, 11311–11316.
24. Brown, I. D. *Acta Cryst.* **1977**, *B33*, 1305–1310.

25. Rutherford, J. S. *Acta Cryst.* **1990**, *B46*, 289–292.
26. O'Keeffe, M. *Acta Cryst.* **1990**, *A46*, 138–142.
27. Brown, I. D.; Shannon, R. D. *Acta Cryst.* **1973**, *A29*, 266–282.
28. Bürgi, H.-B.; Dunitz, J. D. *J. Am. Chem. Soc.* **1987**, *109*, 2924–2926.
29. Grabowski, S. J. *Croat. Chem. Acta* **1988**, *61*, 815.
30. Lippincott, E. R.; Schroeder, R. *J. Chem. Phys.* **1955**, *23*, 1099.
31. Schroeder, R.; Lippincott, E. R. *J. Phys. Chem.* **1957**, *61*, 921.
32. Grabowski, S. J. *J. Mol. Struct.* **2000**, *552*, 153.
33. Steiner, T.; Saenger, W. *Acta Cryst.* **1994**, *B50*, 348.
34. Bertolasi, V.; Gilli, P.; Ferretti, V.; Gilli, G. *Chem. Eur. J.* **1996**, *2*, 925.
35. Alig, H.; Lösel, J.; Trömel, M. *Z. Kristallogr.* **1994**, *209*, 18.
36. Bader, R. F. W. *Atoms in Molecules. A Quantum Theory*; Oxford University Press: New York, **1990**.
37. Cioslowski, J.; Mixon, S. T. *J. Am. Chem. Soc.* **1992**, *114*, 4382–4387.
38. Cioslowski, J.; Surján, P. R. *J. Mol. Struct. (Theochem)* **1992**, *255*, 9–33.
39. Howard, S. T.; Krygowski, T. M. *Can. J. Chem.* **1997**, *75*, 1174.
40. Alkorta, I.; Rozas, I.; Elguero, J. *Structural Chemistry* **1998**, *9*, 243–247.
41. Alkorta, I.; Rozas, I.; Elguero, J. *J. Mol. Struct. (Theochem)* **1998**, *452*, 227–232.
42. Pais, A. A. C. C.; Arnaut, L. G.; Formosinho, J. S. *J. Chem. Soc. Perkin Trans. 2* **1998**, 2577–2584.
43. Dziembowska, T.; Szczodrowska, B.; Krygowski, T. M.; Grabowski, S. J. *J. Phys. Org. Chem.* **1994**, *7*, 142–146.
44. Allen, F. H.; Bellard, S.; Brice, M. D.; Cartwright, B. A.; Doubleday, A.; Higgs, H.; Hummelink, T. W. A.; Hummelink-Peters, B. G. M. C.; Kennard, O.; Motherwell, W. D. S.; Rodgers, J. R.; Watson, D. G. *Acta Cryst.* **1979**, *B35*, 2331.
45. Allen, F. H.; Davies, J. E.; Galloy, J. E.; Johnson, J. J.; Kennard, O.; Macrae, C. F.; Mitchell, E. M.; Mitchell, G. F.; Smith, J. M.; Watson, D. G. *J. Chem. Inf. Comput. Sci.* **1991**, *31*, 187.
46. Krygowski, T. M.; Kalinowski, M. K.; Turowska-Tyrk, I.; Hilberty, P. C.; Milart, P.; Silvestro, A.; Topsom, R. D.; Daehne, S. *Struct. Chem.* **1991**, *2*, 71.
47. Hendrickson, J. B.; Cram, D. J.; Hammond, G. S.; *Organic Chemistry*, 3rd Ed., McGraw-Hill and Kogakusha Ltd, New York and Tokyo, **1970**, p. 296.
48. Hiberty, P. C.; Ohanessian, G. *J. Am. Chem. Soc.* **1984**, *106*, 6963.
49. Maurin, J.; Krygowski, T. M. *J. Mol. Struct.* **1988**, *172*, 413.
50. Talberg, H. J. *Acta Chem. Scand.* **1977**, *A31*, 37.
51. Talberg, H. J. *Acta Chem. Scand.* **1975**, *A29*, 919.
52. Grabowski, S. J.; Krygowski, T. M. *Tetrahedron* **1998**, *54*, 5683–5694.
53. Steiner, T. *J. Chem. Soc. Perkin Trans. 2* **1995**, 1315–1319.
54. Steiner, T. *Cryst. Rev.* **1996**, *6*, 1–57.
55. Desiraju, G. R. *Acc. Chem. Res.* **1996**, *29*, 441–449.
56. Jeffrey, G. A.; Saenger, W. *Hydrogen Bonding in Biological Structures*; Springer: Berlin, **1991**, p. 51.
57. Wessel, J.; Lee, J. C.; Peris, E.; Yap, G. P. A.; Fortin, J. B.; Ricci, J. S.; Sini, G.; Albinati, A.; Koetzle, T. F.; Eisenstein, O.; Rheingold, A. L.; Crabtree, R. H. *Angew. Chem. Int. Ed. Engl.* **1995**, *34*, 2507–2509.
58. Crabtree, R. H.; Siegbahn, P. E. M.; Eisenstein, O.; Rheingold, A. L.; Koetzle, T. F.; *Acc. Chem. Res.* **1996**, *29*, 348–354.
59. Bosque, R.; Maseras, F.; Eisenstein, O.; Patel, B. P. P.; Yao, W.; Crabtree, R. H. *Inorg. Chem.* **1997**, *36*, 5505–5511.
60. Liu, Q.; Hoffmann, R. *J. Am. Chem. Soc.* **1995**, *117*, 10108.
61. Remko, M. *Molecular Physics* **1998**, *94*, 839–842.

62. Alkorta, I.; Elguero, J.; Foces-Foces, C. *Chem. Commun.* **1996**, 1633–1634.
63. Grabowski, S. J. *Chem. Phys. Letters* **1999**, *312*, 542–547.
64. Orlova, G.; Scheiner, S. *J. Phys. Chem. A* **1998**, *102*, 260–269.
65. Orlova, G.; Scheiner, S. *J. Phys. Chem. A* **1998**, *102*, 4813–4818.
66. Orlova, G.; Scheiner, S. *Organometallics* **1998**, *17*, 4362–4367.
67. Orlova, G.; Scheiner, S.; Kar, T. *J. Phys. Chem. A* **1999**, *103*, 514–520.
68. Auf der Heyde, T. P. E.; Bürgi, H.-B. *Inorg. Chem.* **1989**, *28*, 3960–3969.
69. Cieplak, A. S. Organic Addition and Elimination Reactions; Transformation Paths of Carbonyl Derivatives, In *Structure Correlation*, Bürgi; H-B.; Dunitz, J. D. Eds.; Weinheim, **1994**.
70. Auf der Heyde, T. Ligand Rearrangement and Substitution Reactions of Transition Metal Complexes, In *Structure Correlation*, Bürgi, H-B.; Dunitz, J. D., Eds.; Weinheim, **1994**.
71. Bürgi, H.-B.; Shklover, V. Reaction Paths for Nucleophilic Substitution (S_N2) Reactions, In *Structure Correlation*, Bürgi, H-B; Dunitz, J. D.; Eds.; Weinheim, **1994**.
72. Grabowski, S. J.; Krygowski, T. M. *Chem. Phys. Letters* **1999**, *305*, 247–250.
73. Kim, Y. *J. Am. Chem. Soc.* **1996**, *118*, 1522–1528.
74. Neumann, M. A.; Craciun, S.; Corval, A.; Johnson, M. R.; Horsewill, A. J.; Benderskii, V. A.; Trommsdorff, H. P. *Ber. Bunsenges. Phys. Chem.* **1998**, *102*, 325–334.
75. Garcia-Viloca, M.; González-Lafont, A.; Lluch, J. M. *J. Am. Chem. Soc.* **1997**, *119*, 1081–1086 and references cited therein.
76. Smallwood, C. J.; McAllister, M. A. *J. Am. Chem. Soc.* **1997**, *119*, 11277–11281.
77. Robertson, J. M.; Ubbelohde, A. R. *Proc. Royal Society of London A*, **1939**, *170*, 222–240.
78. Ubbelohde, A. R.; Gallagher, K. J. *Acta Cryst.* **1955**, *8*, 71–83.
79. Benedict, H.; Limbach, H. H.; Wehlan, M.; Fehlhammer, W. P.; Golubev, N. S.; Janoschek, R. *J. Am. Chem. Soc.* **1998**, *120*, 2939–2950.
80. Smirnov, S. S.; Benedict, H.; Golubev, N. S.; Denisov, G. S.; Kreevoy, M. M.; Schowen, L. R.; Limbach, H. H. *Can. J. Chem.* **1999**, *77*, 943–949.
81. Brown, I. D. *J. Chem. Soc. Dalton Trans.* **1980**, 1118–1123.
82. Donohue, J. *J. Phys. Chem.* **1952**, *56*, 502–510.
83. Etter, M. C. *Acc. Chem. Res.* **1990**, *23*, 120–126.
84. Taylor, R.; Kennard, O. *J. Am. Chem. Soc.* **1982**, *104*, 5063–5070.
85. Brown, I. D.; Skowron, A. *J. Am. Chem. Soc.* **1990**, *112*, 3401–3403.
86. Brown, I. D. *Acta Cryst.* **1988**, *B44*, 545–553.
87. Kuchitsu, K.; Fukuyama, T.; Morimo, Y. *J. Mol. Struct.* **1969**, *4*, 41.
88. McWeeny, R. *Coulson's Valence*; Oxford University Press, **1979**.

THE PREDICTION OF DENSITY

Guy Piacenza, Guy Jacob and Roger Gallo

1. INTRODUCTION

Density or weight/unit volume (w/v) is expressed as grams/cubic centimeters for solids and liquids and usually as grams/liter for gases.[1]

Density is an important parameter in chemistry and chemical engineering. It is useful in fields as different as estimation of energetic materials

Advances in Quantitative Structure Property Relationships
Volume 3, pages 67–108.

performances,[2–5] structure of coal,[6] properties of polymers,[7,8] or calculations of unit operations[9,10] in chemical processes.

In this review, we will deal only with density of condensed phase because density of gases is readily deduced from molecular weight when the gas is ideal. In domains of temperature and pressure far away from standard conditions, corrective terms are applied to the virial equation. This modification is concerned with thermodynamics and fluid mechanics and will not be treated here.

The experimental measurement of density can be made by several methods. The pycnometric method uses a simple apparatus and is accurate enough to satisfy any technical requirements.[10] Therefore the question can be asked of the interest of the prediction of density since the needed value can often be found in reference texts, in the original literature or in supplier's catalogue (besides mp or bp and n_D^{20}). However, there are some compounds reported in the literature for which no density is given. Sometimes the values of density have been determined experimentally but the data are not easily available. In addition, for all new compounds the density may be of interest for predicting some chemical properties or the physical behavior in synthesis, purification and disposal processes.

Several methods for predicting the density have been proposed with time. As expected they have followed the general understanding of chemical structures and properties; interestingly and not surprisingly they are related more or less to general approaches used for determining other physicochemical parameters like melting points,[11a] enthalpies of formation,[11b,32] or refractive indexes.[39] Early equations for density evaluation were based on elemental analysis or on molecular weight. Later, incremental methods were proposed which used group contributions or van der Waals radii. Recent procedures which include molecular volumes and crystalline structure determination have given more accurate results while treating almost all functional groups.

It is this recent progress, which deserves a detailed examination, that we will mostly discuss here. However simple empirical procedures are still valid for preliminary evaluation. The difficulty is that if papers report and discuss individual predictive methods, no review paper, or chapter of any book, has been so far devoted to a general presentation and critical evaluation of the methods used for the prediction of density. Therefore in this chapter, we will try not only to describe the basic assumptions and the procedures used for each type of calculation method, but also their scope and limitations.

II. EMPIRICAL METHODS

All these methods are based on molar volumes defined as $V_m = \text{mol.wt}/\rho$. (with mol.wt = molar weight and ρ = density). The molar volume of a compound is calculated by the addition of increments, each of which corresponds to a given structural element. According to the nature of the structural element two methods can be distinguished: atomic increments procedures, and group increments procedures.

A. Atomic Increment Procedures

These methods were proposed many years ago. The early report was by Kopp[12]. The calculation is based on elemental analysis. For a compound having the general formula $C_xH_yO_z$, the molar volume V_m is obtained by summing the atomic molar volumes according to:

$$V_m = 11x + 5.5y + 12.2z \tag{1}$$

The density is simply obtained by the Eq. 2

$$D = W/V_m \tag{2}$$

Traube[13a,b] and Le Bas[14a,b] used the same type of linear equation and proposed an extended series of parameters (Table 1) given for fitting a larger number of functional derivatives.

Table 1. Atomic increments used for molar volumes calculations by Traube and Le Bas

Atom	*Atomic increment*[a]	
	Traube	*Le Bas*
C	9.9	14.8
H	3.1	3.7
O (in ethers)	5.5	11.0
O (in C=O)	5.5	7.4
S	15.5	25.6
F	5.5	8.7
Cl	13.2	21.6
Br	13.2	27.0
I	13.2	37.0
Residual volume	26.0	

Note: [a] Atomic increments in cm^3/g-atom.

There is an interesting difference between the methods published by Traube and by Le Bas. For a given compound, Traube added to the sum of the atomic increments a constant value which was called the residual volume. As this was not done by Le Bas, his atomic increments were always larger than the corresponding values of Traube, as appears in Table 1. The method of Traube was used for the calculation of densities of coal constituents.[6] Compared to the values of Le Bas, Traube's values are in much better agreement with those obtained by later investigators like Davis[15] and Harrison[16] who published additional atomic increment values.

A recent improved methodology, based on atomic increments and additional corrections, has been proposed by Girolami.[17a] He first calculates the sum (denoted V_s) of scaled atomic volumes, which are the same for all elements in each line of Mendeleev table (Table 2).

For example, for a molecule having a formula $C_xH_yCl_z$, V_s writes:

$$V_s = 2x + y + 4z$$

Then the density of a given liquid can be calculated from the following simple equation:

$$d = \frac{W}{V_s} \tag{3}$$

where the factor 5 allows the density to be expressed in units of $g\,cm^{-3}$.

Some classes of liquids have densities that are constantly higher than those predicted by Eq. 3 (among these are molecules that have hydrogen-bonding groups or rings).

More precise densities for such liquids can be obtained if the density calculated from Eq. 3 is multiplied by 1.10 for each of the following functional group present in the molecule:

- a hydroxyl group
- a carboxylic acid group

Table 2. Scaled atomic volumes for Girolami's equation

Element	*Scaled atomic volume*
H	1
1st short period; Li—F	2
2nd short period; Na—Cl	4
1st long period; K—Br	5
2nd long period; Rb—I	7.5
3rd long period; Cs—Bi	9

- a primary and secondary amino group
- an amide group (including *N*-substituted amides)
- a sulfoxyde group
- an unfused ring.

For fused rings the density should be multiplied by 1.075 per fused ring (the density of naphthalene derivatives should be multiplied by 1.15). If a molecule contains several of the structural features in the list above, the density should be multiplied by a factor of no more than 1.30. Densities are calculated for a series of 166 organic liquids taken among 180 organic solvents from the reference book "The Chemist's Companion";[17b] (the solids (mp > 30 °C) and gases (bp < 40 °C) were excluded from consideration). Plotting calculated densities (d_{calc}) versus experimental densities (d_{exp}) gives:

$$d_{calc} = 1.01\, d_{exp} - 0.0006 \qquad (4)$$
$$n = 166 \qquad R^2 = 0.982$$

Application of Eq. 3 to a limited series of 23 inorganic liquids gives results that are accurate to within 7% for nearly all of these.

The Girolami's equation brings some improvements to the previous atomic increment equations not only because corrective terms for specific structural features are added but also, more theoretically, because the scaled atomic volumes are derived from tables of van der Waals radii:

Girolami made the simplification of equaling each van der Waals radii[18] of C, N, O, F to the average of N and O vdW radii.

Computing the relative volumes of these elements gives the results reported in Table 3.

The shortcoming, however, is that the scaled atomic value is the same for all elements of the same line. Moreover (this is true for all atomic increments procedures), the molecular formula of the liquid is what is considered, not the actual molecular structure. Indeed the use of atomic increments proved to be too simplistic unless different increments were used for the same atom having different hybridization and geometry. This is the basis of group increment methodology described in the next paragraph.

Table 3. Calculation of scaled atomic increments

Element	r_{vdW}	$4/3\pi(r_{vdw})^3$	*Scaled atomic volume*
H	1.15	6.40	1
1st short period	1.45	12.8	2

B. Group Increment Procedures

The group increment procedures are based on the general principle of additivity. What is considered here is not the additivity, which is involved in many quantitative structure-reactivity relationships, like those of the extension of the Hammett's and Taft's equations to more than only one substituent parameter[19,20] e.g. Eqs. 5 and 6.

$$\log k = \rho_I \sigma_I + \rho_R \sigma_R + \log k_0 \tag{5}$$

$$\log k = \rho\sigma + \delta E_s + \log k_0 \tag{6}$$

These are directed properties related generally to chemical rates and equilibria and to their extension to QSAR.[21,22]

Density, like other physical characteristics, depends on the molecular structure as a whole. Therefore, the general additivity relationship to be considered in this case may be written as:

$$A_i = \Sigma n_{ij} a_j$$

where a_j are the empirical increment of groups (more generally of structural features) and n_{ij} is the number of respective groups in the given structure.[23]

The major early important contribution to the use of group increment procedures for the prediction of density is due to Exner. He first ruled out the use of Parachor which was proposed by Sudgen[24] as an additive increment for physical properties and was used for solving structural problems in chemistry[25,26] and biochemistry.[27] The Parachor (or "comparative volume", from the greek para = by the side and chor = volume) is defined in Eq. 7 where M is the molecular weight, d the density and γ the surface tension.

$$P = Md^{-1}\gamma^{1/4} = \Sigma P_i \tag{7}$$

The $\gamma^{1/4}$ factor (derived from the MacLeod equation) was supposed to allow comparison of molecular volumes at equal surface tension and therefore P was expected to be a true measure of the molecular volume.

Using Eq. 8.

$$\gamma^{1/4}d^{-1} = \frac{C}{M} + \frac{P(\mathrm{CH_2})}{14.02} \tag{8}$$

(where C is a constant characteristic of each homologous series and P (CH_2) is the increment of the methylene group), Exner plotted $\gamma^{1/4}\, d^{-1}$ versus $1/M$ for 13 series of functional derivatives and showed clearly that Eq. 7 is not fulfilled since the lines for each series do not intersect in one point; and consequently the group CH_2 cannot be considered common for all homologues series. Exner concluded therefore that the conception of Parachor has no material significance and that its use should be discontinued.[28]

Then Exner examined the additive character of the molar volume as the simplest additive property of liquids. He expressed the fundamental additive relation as Eq. 10

$$Md^{-1} = \Sigma V_i \tag{10}$$

where V_i represents the increments for the individual structural group and the summation involves the entire molecule.

The validity of this approach is tested by linear correlation using Eq. 11.

$$d^{-1} = \frac{vh}{mh} + \frac{mh\,ve - mc\,vh}{mh} \cdot M^{-1} \tag{11}$$

where vh and vc are respectively the increments for the homologous incremental group and for the constant part of the molecule; mh and mc are the corresponding molecular weights. Plotting d^{-1} versus M^{-1}, Exner showed that linear relations are obtained in all series. They intersect the ordinates at one point (or nearly so) at 20 °C whereas they give a scattering at boiling point indicating that the additivity is better fulfilled at a constant temperature (20 °C) rather than at the boiling point of each individual substance. The accuracy of the additive relation was tested on 17 series of compounds; the sum of which is 870 liquids.[29] From the slope and intercept of the straight lines (obtained by plotting $1/d$ versus $1/M$) the constituent volumes in the homologous series are deduced. When values for these volumes are known, a different homologous series can be selected and values for other constituent volumes can be deduced, and so forth. The values of additive increments obtained by this study are reported in Table 4.

The prediction of the density in the homologous series of non-branched primary functional derivatives is good (standard deviation of *c.* 0.003). The error is larger with non adjacent difunctional derivatives, 1,1-bisderivatives, *ortho*-derivatives and compounds with a single tertiary carbon atom. The additive scheme can be considered as not fulfilled in substances with highly branched aliphatic chains, functional groups directly bonded to each other and for the first member of the homologous series (methyl derivatives). In summary[29] the additive relation (which was applied only to liquids) is useful for the calculation of density of higher homologues on the basis of the lower ones, in the same series, and it is less accurate for polyfunctional derivatives on the basis of monofunctional derivatives.

Table 4. Values of additive increments for the molar volumes derived from primary monofunctional derivatives (at 20 °C)

Group	*Increment (v)*	*Group*	*Increment*
CH_2	16.58	COO	19.50
CH_3	31.48	SH	25.81
$CH_2{=}CH{-}$	42.48	S	10.78
$CH{\equiv}C{-}$	33.22	—S—S—	27.50
Cyclopentyl	79.17	NH_2 (prim amines)	17.67
Cyclohexyl	93.92	NH (sec amines)	7.30
C_6H_5	74.65	N (tert amines)	5.09
1-naphtyl	107.32	CN	22.67
2-naphtyl	109.52	NO_2	24.51
F (primary)	15.11	PO_4	27.43
Cl (primary)	22.96	2-pyridyl	67.66
Br (primary)	26.19	3-pyridyl	66.46
I (primary)	32.93	4-pyridyl	66.35
OH (in RCH_2OH)	10.25	2-$C_5H_4NCO_2$—	87.62
O (in ethers)	6.74	3-$C_5H_4NCO_2$—	88.89
—OCH_2O—	29.11	4-$C_5H_4NCO_2$—	89.55
C_6H_5O—	79.20	2-thienyl	65.58
CHO	25.08	2-C_4H_3SCO—	76.13
CO (in ketones)	10.21	2-$C_4H_3SCO_2$—	87.49
CO_2H	27.24		
H	14.90	C	13.22
Double bond	24.22	five membered ring	11.17
Triple bond	44.76	six membered ring	9.34

Nielsen, extending Exner's method, calculated the densities of fuels and explosives with molar volumes increments, adding constituent volumes related to multi-membered ring systems, and chemical entities found in explosive compounds.[30] He observed that for most studied compounds density is calculated with good accuracy (mean error less than 5%). He found, however, that the first member of the homologous series sometimes deviates from the additive scheme; and that gem and vicinal substitution of electronegative groups cause positive deviations of the calculated values. Strain within polycyclic ring systems (e.g. cubane; [3.3.0] bicyclooctane) lead to low calculated densities of up to 10% deviation.

He suggests that, for these structural parameters causing deviations from the additive scheme, qualitative adjustments in calculated values can be deduced, but he gives no precise quantitative measure of the specific correction to be applied in each specific case. Moreover, the compounds whose densities have been predicted in this study are liquids (mostly) and solids; the calculations have been made in each case without correction of the physical state (liquid or solid) of the compound which in view of later studies discussed in Chap. IV is clearly a shortcoming.

Tarver[31] defined a set of constituent volumes different from those of Exner or Nielsen, and used them according to the same group additivity approach.

The group increments are those of carbon atom (with different substituents) of the framework of the molecule. Values of five of these group increments are based on 25 aromatic compounds but values of the remaining 73 group increments are based on only 148 compounds. Examples of group increments for solid aromatic compounds and for aliphatic compounds are given in Tables 5 and 6 respectively. Examples of density estimations are given in Table 7.

The results of these calculations are good. The largest errors occurred for compounds with bulky groups in close proximity and for compounds that contain group increments whose calculated molar volumes are based on density data for only two or three compounds. Tarver notes a fact observed by the authors of this chapter during studies reported in paragraph IV on Molecular Mechanics Calculations that "some of the apparent discrepancies between predicted and measured values represent flaw in certain measured data rather than faults in the predictive scheme". He suggests the possibility of including second-order effects (like nearest-neighbor interactions, phase change and crystalline structure changes) in a more sophisticated group additivity approach similar to the detailed methods developped by Benson[32,33] for the estimation of various thermo-chemical properties,[34] but he gives no such corrections. In addition, Tarver, like Nielsen, made no differences between liquids and solids in the compounds that made up his basis sets; which is a source of discrepancies and a limitation in the prediction of densities of new compounds, because two different values are expected whether the compound is a solid or a liquid.

Table 5. Group increments for the calculation of densities of solid aromatic compounds

Group increments[a]	*molar volume* ($cm^3\ mol^{-1}$)
C_a—H	11.876
C_a—NO2	29.697
C_a—OH	16.019
C_a—CH_3	25.963
C_a—NH_2	15.663

Note: [a]C_a designates an aromatic carbon atom.

Table 6. A selection of group increments for aliphatic compounds

Group increment	*Letter designation*	*Molar volume* ($cm^3\ mol^{-1}$)
C—H, H_3	a	30.68
C—C, H_2, NO_2	b	42.61
C—C_2, H_2	c	15.69
C—C_2, H, NO_2	d	29.16
C—C, $(NO_2)_3$	e	74.86
C—C, H, $(NO_2)_2$	f	59.12
C—C2, $(NO_2)_2$	g	40.87
C—C, O, =O	h	21.88
C—C, H_2, O	i	16.84
C—C, H_2, OH	j	28.00
C—C, H_2, NH_2	α	34.42
C—C_2, H, NH_2	β	22.49
C—C, H_2, NH	γ	20.64
C—C_2, H, NH	δ	8.42
$(C—C_2, H_2)_c$©	ca	18.38
$(C—H_2, (N—NO_2)_2)_c$	cb	41.60
$(C—C, H_2, O)_c$	cc	21.75
$(C—C, H, =C)_c$	cd	15.12
$(C—H, O, =C)_c$	ce	20.66
$(C—H, =C, NH)_c$	cf	19.81

Note: © subscript c denotes a group from a cyclic compound containing a five- or six-membered ring.

Eremenko,[35] in a series of papers published between 1968 and 1982, has proposed a completely different type of equation relating the density to the percent in mass (by weight) of hydrogen in the molecule (*H*).

$$d = a + kH \tag{12}$$

in which *k* and *A* are empirical coefficients representing the slope and ordinate obtained when plotting *d* versus *H* for an homologous series.

He first studied 35 solid *n*-alkanes (derivatives ranging from $C_{21}H_{44}$ to $C_{29}H_{60}$) obtaining Eq. 13

$$d_4^{20} = 2.2584 - 0.0886H \tag{13}$$

Table 7. Examples of density estimations using group increments of Tables 5 and 6

Compound[a]	*Groups present*[b]	*Calc molar volume*	*Calc density*	*Exp density*	*Error*
o-dinitrobenzene	4(Ca—H) + 2(Ca—NO_2)	106.90	1.572	1.562	0.44%
p-dinitrobenzene	4(Ca—H) + 2(Ca—NO_2)	106.90	1.572	1.625	3.26%
3,5-dinitrophenol	3(Ca—H) + 2(Ca—NO_2) + (Ca—OH)	111.04	1.658	1.702	2.58%
picric acid	2(Ca—H) + 3(Ca—NO_2) + (Ca—OH)	128.86	1.778	1.763	0.85%
2,4-dinitroaniline	3(Ca—H) + 2(Ca—NO_2) + (Ca—NH_2)	110.69	1.654	1.615	2.41%
$CH_3CH_2NO_2$	a + b	73.29	1.024	1.045	1.98%
$CH_3(CH_2)_3NO_2$	a + b + 2c	104.67	0.985	0.971	1.46%
$CH_3CH(NO_2)CH_3$	2a + d	90.52	0.984	0.988	0.38%
$(O_2N)_3CC(NO_2)_3$	2e	149.72	2.004	1.998	0.30%
$CH_3CH(NO_2)_2$	a + f	89.80	1.337	1.350	0.97%
$CH_3C(NO_2)_2CH_3$	2a + g	102.23	1.312	1.300	0.90%
$HOCH_2CH_2NO_2$	b + j	70.61	1.290	1.270	1.56%
$H_2NCH_2CH_2NH_2$	2α	68.84	0.873	0.899	2.93%
$CH_3CH(NH_2)CH_2NH_2$	a + α + β	87.59	0.846	0.858	1.41%
$H_2N(CH_2)_3NH_2$	c + 2α	84.58	0.877	0.884	0.80%
$CH_3(CH_2)_3NH_2$	a + 2c + α	96.48	0.758	0.741	2.25%
$(C_4H_9)_2NH$	2a + 4c + 2γ	165.40	0.781	0.767	1.88%
cyclohexane	6ca	110.28	0.763	0.778	1.98%

Notes: [a] A selection of 18 compounds out of a series of 188 reported in ref. 31.
[b] According to Tables 5 and 6.

For liquid *n*-alkanes, the dependence of the density upon the hydrogen% is also linear (Eq. 14)

$$d_4^{20} = 2.1138 - 0.0837H \tag{14}$$

Excellent linear relations are also obtained for liquid isoalkanes and alkynes (Table 8).

For alkenes and cycloalkanes the equation $d = A + kH$ is unascertainable (in this case there is a general formula C_nH_{2n} giving a constant value $H = 0.1428$). However for alkenes a linear relationship is obtained taking

$$d_4^{20} = 0.8423 - 0.00526P \tag{15}$$

where P is the % by weight of the $CH_2{=}CH$ group whereas for cycloalkanes the density can be expressed by the equation:

$$d_4^{20} = 0.9696 - 0.01121M \tag{16}$$

where M is the % weight of the CH_2 group in the cycloalkane.

Eremenko considered then, following Coulson,[36] the specific atomic volume V_{sp} expressed as:

$$V_{sp} = \frac{4\pi r^3}{3A} \tag{17}$$

(where r is the covalent radius and A the atomic weight of the atom). Since V_{sp} is nearly constant for hydrogen, nitrogen, oxygen and chloride (Table 9), he extended his relation to series of homologous molecules with a large variety of functionalities[37] (Table 10).

For 1-alkanols and 2-alkanones (like for alkenes and cycloalkanes) the equation $d_4^{20} = A + kH$ is insolvable (the general formula $C_nH_{2n}O$ gives a constant value of $H = 0.1282$). A simple alternative equation is:

$$d_4^{20} = B + kX \tag{17a}$$

Table 8. Parameters A and k, from the equation $d_4^{20} = A + kH$, for hydrocarbon series

Series	*Number of compounds*	*A*	*k*
2-methylalkanes	7	2.1260	−0.08987
2,2-dimethylalkanes	5	2.1818	−0.09361
1-alkynes	8	0.0200	−0.0570

Table 9. Covalent radii and specific atomic volumes according to Coulson

Atom	*Covalent radius* (A°)	$V_{sp}(A^{\circ 3})$
Hydrogen	0.30	0.112
Nitrogen	0.73	0.116
Oxygen	0.74	0.106
Chlorine	0.99	0.118
Tin	1.40	0.097

Table 10. Parameters A and k of the equation $d_4^{20} = A + kH$ for series of functional derivatives

Series	*Number of compounds*	*A*	*k*
n-alkanoic acids	9	1.1918	−0.0250
n-alkyl esters of *n*-alkanoic acids	19	0.9967	−0.0011
crystalline α, ω-dicarboxylic acids	8	1.6720	−0.049
n-alkylamines	5	2.6464	−0.1257
nitriles of *n*-carboxylic acids	9	0.6815	+0.0113
liquid nitro-alkanes	8	1.1918	−0.0250
amides	7	1.1605	−0.0127
hydrazides of *n*-alkanoic acids	5	1.5339	−0.0391

where X is the % by weight of oxygen, and B and k are empirical parameters (Table 11).

Eremenko also found a linear relationship between density and refractive index (n_D^T):

$$d_4^{20} = A_1 + k1\, n_D^{20} \tag{18}$$

for alkanes and alkenes[38] and for a series of 15 homologous series of different functional derivatives.[39] Later he described density calculations of

Table 11. Parameters B and k of the equation $d_4^{20} = B + kX$ for alcohols and ketones

Series	*Number of compounds*	*B*	*k*
1-alkanols	9	0.8424	−0.0015
2-alkanones	6	0.8400	−0.00157

aromatic crystals[40] using Eq. 12, and found linear relationships (Table 12) for polycondensed aromatic (Series A) fluoro and chlorosubstited aromatics (Series B) and substituted benzoic acids (Series C).

He noticed that this method for calculating densities of aromatic compounds has a major limitation: it does not make the difference between isomeric and polymorphous compounds.

In a later paper on interrelation between density and structure of energetic materials,[41] he gave further examples of this behavior (couples 1–2 and 3–4 in Table 13).

Consequently he made predictions on two different series of liquid derivatives: "symmetrical" (mostly α- ω- dinitro- and dinitrato-derivatives) and asymmetrical (all other isomeric nitro- and nitrato-compounds).

For symmetrical derivatives the correlation equation is:

$$d_4^{20} = 1.780 - 0.096H \qquad (19)$$
$$n = 9 \qquad R = 0.998$$

For asymmetrical derivatives the correlation equation is:

$$d_4^{20} = 1.584 - 0.067H \qquad (20)$$
$$n = 34 \qquad R = 0.98$$

Table 12. Coefficient of the correlation $d_4^{20} = A + kH$ for aromatic series

Series	*Number of derivatives*	*A*	*k*	*R*
A	6	1.946	−0.127	0.995
B	8	1.982	−0.152	0.997
C	7	1.965	−0.131	0.999

Table 13. Effect of symmetry of liquid nitro derivatives on their density

Number	*Compound*	*H parameter*	*Density*
1	$CH_3CH(NO_2)_2$	3.36	1.364
2	$O_2NCH_2CH_2NO_2$	3.36	1.460
3	$CH_3CH_2CH(NO_2)_2$	4.51	1.258
4	$O_2NCH_2CH_2CH_2NO_2$	4.51	1.354

For solid energetic materials good correlation are also obtained. They are summarized in Table 14.

The methodology used by Eremenko et al. is in-between an atomic increments procedure and a group increments procedure. The parameter H is indeed a percent by weight of hydrogen (from elemental analysis), but the parameters A and k are characteristic of given classes of compounds including the framework (normal alkyl, isoalkyl, aromatic) and the functionality in the series. The advantages of the procedure are that solid and liquid compounds are considered separately and that several parameters A and k have been obtained. The important shortcoming is that calculations can be made only within narrow specific series already studied which make the method useless for new compounds.

Following a different methodology, Stine developed a procedure[42,43a] for predicting density which assumes that the molar volume of a molecule (V) in the unit cell of a crystal (which in this case includes some of the intermolecular volume) is the sum of the volumes of the molecular constituents. The molecular constituent is the atom in its given bonding environment. For example an sp^3 carbon atom with four single bonds is designated C (1, 1, 1, 1), or C(1, 1, −1, −1) if two of the bonds were in a ring system. An sp^2 carbon atom with a double bond and two single bonds is C(2, 1, 1). A negative value indicates that the bond is part of a ring system.

Compounds consisting only of the elements hydrogen, carbon, nitrogen, oxygen and fluorine are considered. The molar volume V is assumed to be a

Table 14. Coefficient of the equation $d_4^{20} = A + kH$ for solid energetic materials

Type of compound	*Number of compounds*	*A*	*k*	*R*
aliphatic compounds containing and(or) nitrate groups	29	2.114	−0.169	0.99
aliphatic secondary polynitroalkylamines and polynitroalkylamides	46	2.114	−0.151	0.99
aliphatic primary nitramines and nitrates of primary amines	13	2.118	−0.103	0.97
nitro- and (or) nitroxycyclanes or oxacyclanes	12	2.085	−0.143	0.97
nitroazacylanes and nitroazaoxacyclanes	6	2.086	−0.093	0.95

linear combination of the constituent volumes v_i:

$$V = \sum_{i=i}^{n} ai\, v_i \qquad (21)$$

The constituent volumes, which were derived by least squares calculations from over 2000 organic molecules with know crystal structures, are reported in Table 15.

The calculated densities are within 3–4% of the observed values.

The Stine's procedure is probably the most frequently applied, specially with energetic materials, because it can be used easily without sophisticated computation.

Table 15. Constituent volumes for the prediction of density by the Stine's method

Constituent volume	*Volume* ($A^{\circ 3}$)	*Density* (g/cm^3)
H(1) bonded to a carbon	5.981	0.278
H(1) bonded to an aromatic carbon	7.499	0.221
H(1) bonded to a nitrogen	5.199	0.319
H(1) bonded to an oxygen	0.366	4.536
C(1,1,1,1)	13.390	1.486
C(1,1,−1,−1)	11.709	1.702
C(1,−1,−1,−1)	9.755	2.042
C(−1,−1,−1,−1)	9.673	2.060
C(2,1,1)	14.565	1.368
C(1,−1,−2)	13.288	1.499
C(2,−1,−1)	12.654	1.575
C(−1,−1,−2)	10.410	1.914
C(1,−5,−5)	12.094	1.647
C(−1,−5,−5)	10.618	1.876
C(−5,−5,−5)	10.370	1.921
C(3,1) or C(2,2)	16.579	1.202
N(2,2,1)	10.368	2.242
N(1,1,1)	9.234	2.517
N(1,−1,−1)	9.798	2.372
N(−1,−1,−1)	8.759	2.654
N(−5,−5)	12.268	1.895
N(2,1)	14.488	1.605
N(−1,−2)	11.941	1.947
N(3)	15.599	1.490
O(1,1)	12.178	2.181
O(−1,−1)	12.172	2.183
O(2)	12.754	2.063

Ammon and Mitchell[43b] have extended the methodology used by Stine. They propose a new parameterization derived from over 11,000 crystal structures which provides 78 atom and group volumes for C,H,N,O and F containing materials. They assume that the volume of a molecule, or V_{VA} ($_{VA}$ denotes "Volume Additivity"), is the linear sum of individual atom and group volumes. They use the observed unit cell volumes per molecule or V_{obs} ($V_{obs} = V_{cell}/Z$; with Z = number of molecules per unit cell) to determine the individual volumes V_j by minimization of ΔL according to Eq. 21a:

$$\Delta L = \sum_{i=1}^{N} [(V_{obs})i - (V_{VA})i]^2 \qquad \text{with} \quad V_{VA} = \sum_{j=1}^{n} (V_j) \tag{21a}$$

where V_j will be called "linear volumes" because V_{VA} is a linear function of the V_j.

An alternative approach is to determine the V_j by minimization of Δ_{NL} (Eq. 21b) which uses the observed and calculated densities (ρ_{obs}, ρ_{VA})

$$\Delta_{NL} = \sum_{i=1}^{N} [(\rho_{obs})i - (\rho_{VA})i]^2 \qquad \text{with} \quad \rho_{VA} = M_{mol}/0.6022\, V_{VA} \tag{21b}$$

In this case the V_j are referred to as "non linear volumes" because ρ_{VA} is a non linear function of V_j's.

Both "linear" and "non linear" volumes where calculated in this investigation as in the study of Stine.[42] On the other hand Immirzi and Perini[64] and Tarver[31] reported only "linear volumes".

The Ammon's procedure (like that of Stine) is easy to use, and according to the authors it gives a precision slightly improved compared to that of Stine.

In summary, the advantage of group increment procedures is that they are easy to use and that they give good predictions. The limitations are that they often lack complete coverage of parameters. That sometimes the available parameters are valid for liquids only, or for solids only, not for both liquids and solids. That they do not discriminate among regioisomers. That they give important errors with gem and vicinal group interactions or with steric strain in highly branched isoalkyl groups or in polycyclic alkenes.

In other words, there was a need of a better fundamental description of the structure of the molecules based on geometrical parameters for the prediction of density. This is the major concern of the studies reported in the two following chapters (Chapter III and Chapter IV).

III. MOLECULAR VOLUMES AND CRYSTAL STRUCTURE

The methodology of empirical methods reported in Chapter II involves the estimation of the molar volume of a compound (in cm^3/mole) or equivalently its crystal volume ($Å^3$/molecule). These volumes are constituted of individual increments, each of which includes a major contribution from the actual volume occupied by the molecule, and a minor part due to the free space between molecules.

Alternatively, the molecular volume of the molecule and the free space between molecules can be estimated separately. This is an approach which started with the pioneering work of Kitaigorodski[44] and which has been carried out since by several authors using computational methods of increased efficiency.

Kitaigorodski noticed that the packing of the molecule in the crystal is determined mostly by interactions between hydrogen atoms or between hydrogen atoms and other elements. The ideal packing is obtained for a minimum of interatomic distances. The molecular packing coefficient (or "packing coefficient"), defined as the ratio of the molecular volume to the crystal volume, assumes scattered values in the range 0.66–0.77 (Gavezotti has described more recently[45,46] extended values of average packing coefficient of different chemical classes). The morphologic change associated to a loss of symmetry results in an increase of the packing coefficient. The change from solid to liquid state implies a sharp decrease of the packing coefficient (e.g. for benzene and naphthalene, the coefficient $k \geq 0.68$ for the solid state drops to 0.58 for the liquid); in addition a supplementary increase of temperature up to the boiling point lowers the packing coefficient (this will be treated in Chapter VI). From these observations two general strategies, for the determination of density, can be anticipated:

(a) calculation of the individual volume of the molecule and prediction of the crystal packing from the structure of its constituent molecules,
(b) calculation of the volume of the molecule and use of an estimated packing coefficient, including corrective terms if necessary.

Several authors have tried to compute the molecular packing in crystals as a part of ab initio prediction of crystal structure.[47,47 bis]

Because most of organic compounds crystallize in several polymorphs, the number of which is inherently unknown, prediction deals with evaluation of relative energies of most stable packings. Isolation of them is up to now left to practical trials.

As detailed explanation of methods is beyond the subject of this book, the reader is referred to original papers.[48–57,59–62] Only main guidelines are outlined below.

Due to large number of atoms in the crystal, calculations are performed by molecular mechanics, with special fittings.

Force fields have been extended to nonbonded atom–atom interactions parameters, keeping form of popular potential functions to describe these interactions:

$$E(R) = A\exp(-BR) - CR^{-6} \quad \text{(Buckingham)}$$
$$E(R) = AR^{-12} - BR^{-6} \quad \text{(Lennard–Jones)}$$

where A, B and C are treated as empirical constants for each type of atom–atom interactions and R is the intermolecular distance between atoms.

van der Waals interactions, which cause large part of crystal energies of polar molecules, are often computed from atomic charges derived from quantum mechanics on isolated molecules.

Special attention has been devoted to limitation of long range interactions by carefully chosen cutoff distances and wide use of Ewald summation.

Programs are written to compute packings in only selected space groups among the 230 ones of crystallographic tables. Indeed, 92% of organic compounds crystallize in 18 space groups.

Minimization of energy is usually achieved either by variation of cell parameters or by Monte Carlo approach followed by a clusterization step of closest structures and final optimization.

Of course, limitation lies in the fact that structures belong to selected space groups. Williams[58] described program mpa which requires only the number Z of molecules in the unit cell and returns symmetry operations. Correct attribution of urea in the rare P-421m is reported.

For our concern, major interest of crystals structure prediction lies in the possibility to examine different polymorphs of the same molecule, which is impossible with all other method of density evaluation.

Interest in crystal structure prediction is now beyond theoretical topic because of applications in different domains where solid state is involved: pigments, pharmaceuticals, electronics, explosives. Economical aspect is so important that MSI (Molecular Simulations, San Diego) commercialize a software module called "Polymorph Predictor", designed by Leusen on the basis of the work of Karfunkel at Ciba-Geigy.[63]

Even if all the above mentioned programs are not able to give directly crystal structure from the sole developed formula and need help of experimental determination of one polymorph, no doubt that improvements in this major field will lead to "true" ab initio prediction in next years.

Due to the present difficulty to deduce, ab initio, the crystal packing from the sole geometry of the constituent molecules, a complementary approach has been made which approximates experimentally the crystal packing (packing coefficient).

For finding a rationalization to crystal packing, Immirzi and Perini[64] have explored the possibility of expressing the specific volume of organic substances in the crystal state as the sum of increments including corrective terms. They calculated crystal volume increments from 500 crystalline organic compounds selected according to the following rules: $50 < MW < 500$, liquid compounds are excluded, crystals containing solvent molecules are excluded. Only the following atoms are considered: H, C, O, N, S, F, I, Br, Cl, Na, K and Rb; only benzene and naphthalene are considered as cyclic compounds; There is a probability of H-bond formation between heteroatoms when the distances O...O, O...N, and N...N are $=<$ 2.7–2.8 Å; coordination numbers are: 4, 3 and 2 for C, 2 and 1 for O and 3, 2 and 1 for N. Unitary crystal volumes (c.v.) thus obtained are compared to Kitaigorodsky's molecular volumes (m.v.) in Table 16.

The ratio m.v./c.v. appears rather variable and smaller for the smaller volumes. A possible explanation is that the excess of volume tends to distribute equally among all atoms giving rise to volume increases proportionally larger for the smaller atoms. It turns out, therefore, that the comparison between m.v. and c.v. does not give a clear and transposable description of the packing coefficient. In addition, the increments c.v. defined by Immirzi and Perini are not basically different of other group increments like those of Nielsen, Tarver or Stine.

Cady, following another approach,[65] tried to calculate the geometry of the molecule and to incorporate, in his equation, a formula designed for taking into account the packing coefficient. He calculates the volume of the molecule as the sum of the volume increment of each atom. Each individual volume increment is expressed by the van der Waals sphere (from Bondi's radii),[18] and corrections are made for the volumes of overlapped segments. Cady obtains a Theoretical Maximum Density (TMD) expressed as:

$$\mathrm{TMD} = \frac{MW}{V} \tag{24}$$

Table 16. Unitary crystal volumes (c.v.) versus Kitaigorodsky's molecular volumes (m.v.)

Increment	*m.v.*	*c.v.*	*Ratio m.v./c.v.*	*Increment*	*m.v.*	*c.v.*	*Ratio m.v./c.v.*
—H	2.0	6.9	0.29	—N	10.0	16.0	0.62
—C≡	5.9	15.3	0.38	—F	9.6	12.8	0.75
—C⊆	8.4	13.7	0.61	—Cl	19.9	26.7	0.75
>C<	5.0	11.0	0.45	—Br	26.6	33.0	0.80
=O	13.1	14.0	0.94	H—Cε	11.1	17.9	0.62
—O—	11.8	9.2	1.28	H—C⊆	14.7	22.2	0.67
—NH_2	19.7	21.0	0.94	>CH_2	17.1	24.8	0.69
—NO_2	23.0	25.6	0.90	—CH_3	23.5	31.7	0.74

and predicts the density by the formula

$$d_{\text{Cady}} = \text{TMD}\ k \tag{25}$$

where the parameter k is supposed to be equivalent to a packing coefficient

$$k = 0.7686 - 0.1280\left(\frac{\text{nb}\ CH}{TA}\right) \tag{26}$$

(where nb CH is the number of hydrogen atoms bonded to carbon atoms, and TA is the total number of atoms in the molecule).

In fact, if Cady introduces geometrical parameters and not molar empirical parameters, he has made a prediction of density from the "2D" structural formula of organic compounds.

In addition, his "packing coefficient" k makes no difference between liquids and solids.

Ideal packing coefficient determination would clearly use true "3D" prediction of crystal structure as already mentioned. The only authors who have described such trials for the specific purpose of density prediction are Cromer, Ammon and Holden.[65b] Their program MOLPAK starts from molecular geometry computed by mean of MM or quantum mechanics; then, coordination sphere is built with other identical molecule taking in account the selected space group symmetry; lastly, final refinement of unit cell is achieved by variation of cell parameters and molecules positions. Density is calculated from the lowest energy arrangement. Accuracy of 2% is reported.

This method goes beyond the one of Cady since the crystal packing is computed and not estimated from number of H atoms in the whole molecule. However, the major drawbacks, mentionned by the authors, are the complexity of the method and the amount of computer time required for exploration of the most probable space groups.

IV. MOLECULAR MECHANICS AND CORRELATION ANALYSIS

A. Density by Molecular Mechanics and Correlation Analysis

A recent approach to the prediction of density combines Molecular Mechanics and correlation analysis.

In general, the density of a liquid or of a solid compound is expressed by:

$$d = \frac{W}{(V_{\mathrm{mol}} + V_{\mathrm{inter}})\mathrm{NA}} \tag{27}$$

Where
W = molecular weight;
V_{mol} = molecular volume of the molecule;
V_{inter} = intermolecular free space between molecules;
NA = Avogadro's number.

Molecular Mechanics will calculate (vide infra) a molecular volume which is overestimated (the intermolecular free space is partly filled due to the overestimation of vdW radii in Molecular Mechanics) and correlation analysis will allow to estimate accurately the remaining intermolecular free space, giving a precise calculation of density, on a wide range of d values (say between 0.6 and 2.8), for both liquid and solid compounds.

Before recent applications to the prediction of density, Molecular Mechanics (MM),[66–68] proved useful in the prediction of conformational geometries. It has been used for modeling organic compounds,[69] organic synthesis[70] or reactivity,[71] inorganic compounds[72,73] or polymers.[74,75] Like for all these studies, Molecular Mechanics applied to density give first an optimized geometry, by minimizing the classical equation of energy:

$$E = E_{\mathrm{l}} + E_{0} + E_{\mathrm{W}} + E_{\mathrm{VdW}} \tag{28}$$

The outcome of the calculation is the skeleton of the molecule (bond lengths and angles) pictured like a Dreiding model or like an "ORTEP" drawing from X ray.

The calculation of the volume of the molecule[76] can then be made by various methods including the dot algorithm of Connoly[77] or the grid or cube methods,[78] or related procedures.[79,80]

Piacenza et al.[81] estimated the molecular volume by using a variant of the cube method.[82]

The molecule (in its MM minimum energy conformation) was placed in a parallelepiped with each atom surrounded by the sphere corresponding to its MM van der Waals volume (whereas previous calculations[76–80] used physical van der Waals volumes mostly from Bondi[18]). Successive and systematic testing for the presence or absence of MM van der Waals volume, in small fractions of the total volume of the parallelepiped,[81] leads to the MM molecular volume = V_{molMM}. Keeping in mind that MM van der Waals radii are larger[3] than physical van der Waals radii (from Bondi[18]), as depicted in Figure 1, the MM molecular volume (V_{molMM}) should contain part of the volume between molecules (V_{inter}). Therefore, the MM density

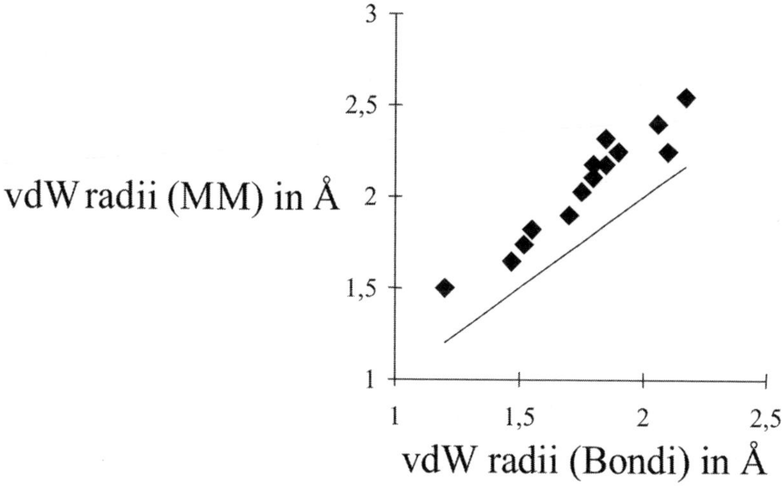

Figure 1. Physical van der Waals radii (Bondi) versus MM van der Waals radii (Allinger).

(d_{MM}) expressed as:

$$d_{MM} = \frac{W}{V_{molMM}NA} \tag{29}$$

should be very close to the actual value of the density (d) which includes the molecular volume (V_{mol}) and the molecular volume (V_{inter}), according to:

$$d = \frac{W}{(V_{mol} + V_{inter})NA} \tag{30}$$

However the d_{MM} values of 80 selected organic compounds (50 liquids and 30 solids) reported in Table 17 show an excellent agreement (a mean error 100 ($|d_{MM} - d_{exp}|/d_{exp}$) of 1.36%) with experimental densities, only for compounds having an experimental density larger than 1.7 (compounds n° 59, 60, 61, 64, 66, 68, 70, 71 and 72 in Table 17). For the other solids the error increases steadily when going to smaller values of density. For all liquids the error is important (an overestimation of 15–30% in all the range of densities). These discrepancies are rationalized by correlation analysis which allows to elaborate a general method of calculation for both solids and liquids and to give an explanation for the behavior of the whole set of data.

Plotting densities calculated by Molecular Mechanics (d_{MM}) versus experimental densities (Figure 2) reveals one line for liquids and one line

Table 17. Experimental and MM densities of 80 selected compounds (50 liquids and 30 liquids)

Compound	*Nature*	d_{exp}	d_{MM}	*Deviation (%)*[a]
hexane	liq.	0.659	0.874	32.63
heptane	liq.	0.683	0.887	29.87
1-pentyne	liq.	0.695	0.925	33.09
diethylamine	liq.	0.707	0.925	30.83
diethyl ether	liq.	0.714	0.963	34.87
1-decene	liq.	0.742	0.932	25.61
dibutylamine	liq.	0.760	0.948	24.74
cyclohexane	liq.	0.779	0.962	23.49
ethanol	liq.	0.789	1.000	26.69
acetone	liq.	0.790	1.030	30.40
methanol	liq.	0.791	1.015	28.25
propanal	liq.	0.806	1.028	27.58
tetradecanoic acid	sol.	0.862	1.048	21.55
cyclohexylamine	liq.	0.867	1.045	20.53
1,2-diaminopropane	liq.	0.874	1.052	20.37
decanoic acid	sol.	0.895	1.073	19.89
ethyl formate	liq.	0.923	1.177	27.52
1-nitrohexane	liq.	0.940	1.126	19.79
butanoic acid	liq.	0.957	1.181	23.41
cyclohexanol	sol.	0.960	1.086	13.13
methyl formate	liq.	0.974	1.230	26.28
benzylamine	liq.	0.981	1.163	18.52
camphor	sol.	0.992	1.146	15.52
acetamide	liq.	0.998	1.242	24.37
water	liq.	1.000	1.234	23.40
propionic anhydride	liq.	1.022	1.228	20.16
(2-methyphenyl)methanol	sol.	1.023	1.187	16.03
benzaldehyde	liq.	1.041	1.244	19.44
acetic acid	liq.	1.050	1.308	24.67
acetic anhydride	liq.	1.081	1.317	21.83
N-acetylethanolamine	sol.	1.108	1.245	12.37
formamide	liq.	1.133	1.356	19.64
nitromethane	liq.	1.137	1.363	19.88
acetamide	sol.	1.159	1.242	7.16
N-acetyl-2-methylaniline	sol.	1.168	1.228	5.14
4-chloroaniline	liq.	1.169	1.436	22.84
benzyl phenyl ketone	sol.	1.201	1.253	4.33
nitrobenzene	liq.	1.204	1.404	16.61
N-phenylacetamide	sol.	1.210	1.255	3.68
methyl nitrate	liq.	1.217	1.492	22.60
acetic acid	sol.	1.266	1.308	3.32

(continued)

Table 17. Continued

Compound	*Nature*	d_{exp}	d_{MM}	*Deviation (%)*[a]
1-bromobutane	liq.	1.276	1.592	24.78
4-acetylamino phenol	sol.	1.293	1.372	6.11
benzoic acid	sol.	1.321	1.372	3.86
dichloromethane	liq.	1.330	1.697	27.59
2-amino-5-nitrotoluene	sol.	1.366	1.396	2.20
4-aminobenzoic acid	sol.	1.374	1.422	3.49
4-chloroaniline	sol.	1.429	1.436	0.49
bromoethane	liq.	1.460	1.895	29.79
maleic anhydride	sol.	1.480	1.611	8.85
nitric acid	liq.	1.504	1.801	19.75
4-chlorobenzoic acid	sol.	1.541	1.562	1.36
2-chloroacetic acid	sol.	1.580	1.644	4.05
trinitromethane	liq.	1.590	1.885	18.55
1-iodobutane	liq.	1.615	2.006	24.18
tetranitromethane	liq.	1.638	2.021	23.38
2-dinitromethylene-1, 3-diazacyclohexane	sol.	1.666	1.541	7.50
1,3,5-trinitrobenzene	sol.	1.680	1.741	3.63
3-methoxy-2,4,6-trinitroaniline	sol.	1.710	1.755	2.63
2,4,6-trinitroaniline	sol.	1.762	1.788	1.48
N,N-difluoro-2,4-dinitroaniline	sol.	1.780	1.812	1.80
1,2-dibromobutane	liq.	1.791	2.132	19.01
sulphuric acid	liq.	1.827	2.017	10.40
2,3,4,6-tetranitroaniline	sol.	1.870	1.876	0.32
pyrosulphonic acid	liq.	1.900	2.199	15.74
2-bromobenzoic acid	sol.	1.929	1.922	0.36
bromochloromethane	liq.	1.934	2.364	22.23
3-trifluoromethyl-2,4,6-trinitroaniline	sol.	1.936	1.986	2.58
bromodichloromethane	liq.	1.980	2.478	25.15
3-bromo-2,4,6-trinitroanisole	sol.	1.987	2.031	2.21
trans-difluorodinitroethylene	sol.	2.048	2.051	0.15
3-bromo-2,4,6-trinitroaniline	sol.	2.157	2.140	0.79
1,2-dibromoethane	liq.	2.180	2.544	16.70
1,4-diiodobutane	liq.	2.358	2.729	15.73
1,2,3-tribromopropane	liq.	2.409	2.723	13.03
dibromochloromethane	liq.	2.450	2.947	20.29
dibromomethane	liq.	2.496	2.929	17.35
tribromonitromethane	liq.	2.793	3.226	15.50
1,1-diiodoethane	liq.	2.840	3.346	17.82
tribromomethane	liq.	2.887	3.361	16.42

Note: [a] Deviation (%) = $(|d_{MM} - d_{exp}|/d_{exp})$ 100

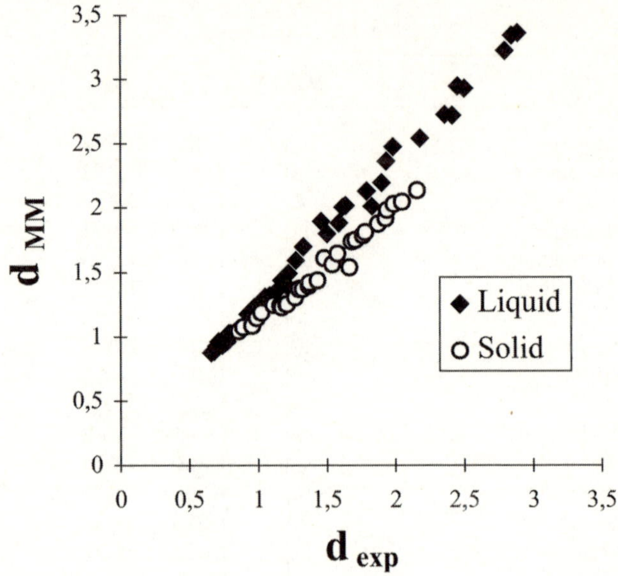

Figure 2. Molecular mechanics estimated density (d_{MM}) versus experimental density (d_{exp}).

for solids. A least square regression gives the corresponding equations:

$$\text{for liquids: } d_{MM} = 1.128 d_{exp} + 0.111 \quad n = 50 \quad R = 0.997 \tag{31}$$

$$\text{for solids: } d_{MM} = 0.873 d_{exp} + 0.244 \quad n = 30 \quad R = 0.990 \tag{32}$$

The quality of the fit can be expressed by a mean deviation (a parameter frequently used in density prediction), taken as:

$$\text{mean deviation} = \frac{1}{N}\sum_{N} \frac{|d_{MM} - d_{\text{regression line}}|}{d_{\text{regression line}}} \tag{33}$$

of 2.37% for liquids and 2.42% for solids.

Given the extended variety of functionalities and of structures (both aliphatic and aromatic), together with the wide range of density values (0.659 to 2.887), the quality of these correlations is surprisingly good. A precise value of calculated density can be obtained from the d_{MM} value;

solving the couple of Eqs. 34 and 35.

$$d_{MM} = ad_{exp} + b \tag{34}$$

$$d_{calc} = d_{exp} \tag{35}$$

gives the simple general Eq. 36

$$d_{calc} = (1/a)d_{MM} - b/a \tag{36}$$

which results in two equations (37 and 38) giving a precise estimation of density (d_{calc}) from the approximate value of density (d_{MM}) obtained directly from the MM molecular volume (V_{molMM}):

$$\text{for liquids:} \quad d_{calc} = 0.0887d_{MM} - 0.098 \tag{37}$$

$$\text{for solids:} \quad d_{calc} = 1.145d_{MM} - 0.279 \tag{38}$$

These two equations were tested for predicting the densities of 183 organic compounds (143 liquids and 40 solids) ranging from 2-methylbutane ($d = 0.620$) to diiodomethane ($d = 3.325$). The calculated densities (d_{calc}) are plotted versus experimental densities in Figure 3.

The correlation line, corresponding to the least square fit is:

$$d_{calc} = 0.995d_{exp} - 0.010 \tag{39}$$

$$n = 183 \qquad R = 0.993$$

The mean error (defined in Eq. 7) is 3.07%. It is comparable to errors obtained with group increments methods when applied within limited series of parent compounds. It is smaller than the same error obtained with incremental methods on enlarged series examined in previous studies.

In addition, correlation analysis gives a deeper insight into the nature of the physical properties of solids and liquids. The difference of experimental density for a compound having a comparable MM molecular volume but being in the liquid or solid state is evident from the two lines of Figure 2. This comes from the well known observation that interatomic distances are larger in liquids than in solids.[44] Stated differently, for most compounds (a major exception is water); the density increases when going from the liquid to the solid state. Values of Table 18 (coming from Eqs. 31 and 32 derived from Figure 2) show that this difference in density (when going from the liquid to the solid) is not constant but increases with the absolute value of density.

However since the molecular volumes used for the above analysis come from Molecular Mechanics, the question of the general character of this behavior must be addressed. For checking this important point, Molecular Mechanics have been used for predicting the density of energetic materials

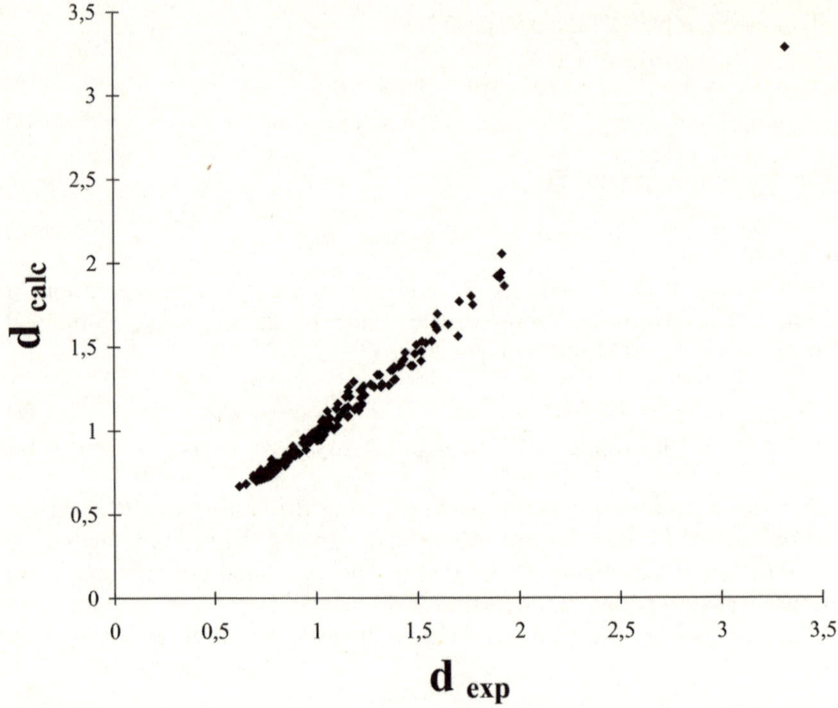

Figure 3. Calculated densities (d_{calc}), derived from d_{MM}, versus experimental densities (d_{exp}).

since these series have been widely studied by group increment methods. In addition, for the same series, the results have been compared to those obtained with group increment parameters and equations of Stine[42a] Holden and Cady[65a] described and discussed in Chapters II and III.

Table 18. Difference of density for compounds in the liquid and solid state

Density of the liquid	*Density of the solid*	*Δd (sol–liq)*[a]
1.	1.13	0.13
1.5	1.78	0.28
2	2.43	0.43

Note: [a]Δd from eqs. 31 and 32.

B. Improvements of Previous Procedures

Molecular Mechanics (MM) were first used for calculating the density of a series of 142 energetic materials (99 solids and 43 liquids) having experimental densities ranging from 1.036 (isopropyl nitrate) to 2.040 (hexanitrohexaazaisowurtzitane or HNIW or CL20). The densities were calculated according to Eqs. 29 and 36 as described above. Plots of d_{MM} versus d_{exp} in Figure 4 show again (like in Figure 2) one line for liquids and one line for solids.

The correlation equations are:

$$\text{for liquids:} \qquad d_{MM} = 1.017 d_{exp} + 0.111 \qquad (40)$$

$$\text{for solids:} \qquad d_{MM} = 0.903 d_{exp} + 0.159 \qquad (41)$$

indicating that the difference in density between the liquid ant the solid is increasing steadily with the absolute value of density.[61]

Then the method for predicting density developed by Stine (described in Chapter III) was applied to the same series of 99 solid and 43 liquid energetic materials. The crystal density (d_{Stine}) being obtained by dividing the molecular weight by the crystal molecular volume (V_{cm})

$$d_{Stine} = M/V_{cm} \qquad (42)$$

Although Stine's method was originally designed for the solid state,

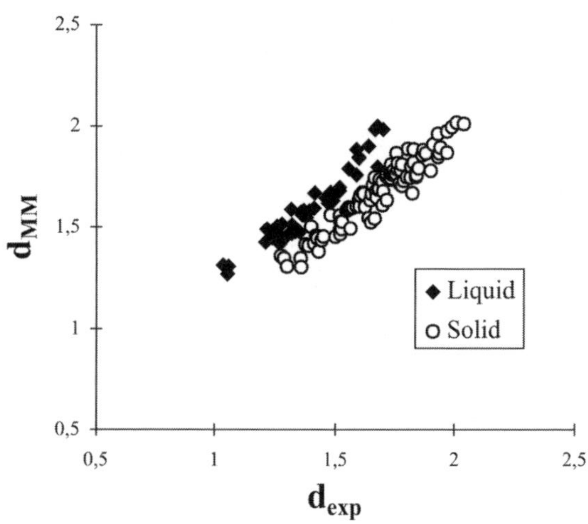

Figure 4. MM estimated density (d_{MM}) versus experimental density (d_{exp}) for 142 energetic materials (99 solids and 43 liquids).

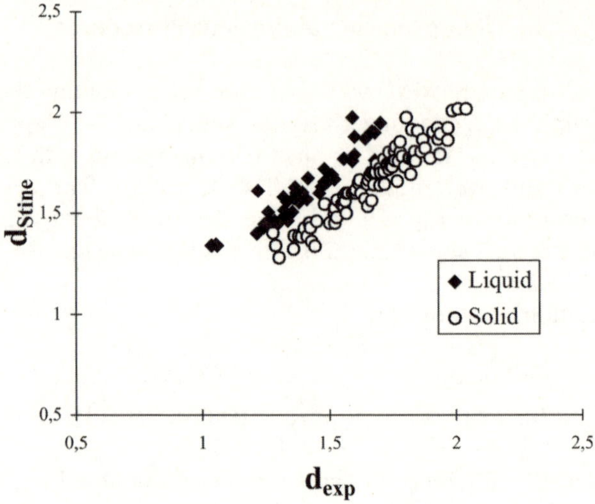

Figure 5. Stine calculated density (d_{Stine}) versus experimental density (d_{exp}).

correlation analysis, once again, allows to analyze both solids and liquids. Indeed the values of d_{Stine} plotted versus experimental densities in Figure 5 show clearly one line for liquids and one line for solids.

This is in agreement with the fact that group increments used by Stine have been obtained from crystals. So that application to liquids should give systematic overestimated values. Indeed correlation equations from least squares fit are:

$$\text{for liquids:} \qquad d_{Stine} = 0.916 d_{exp} + 0.328 \tag{43}$$

$$\text{for solids:} \qquad d_{Stine} = 0.925 d_{exp} + 0.116 \tag{44}$$

The similarity with the Molecular Mechanics calculations is striking. Whereas the two methodologies and parameterization are completely different, Figures 4 and 5 show two almost parallel lines in the 2 cases which are separated by nearly the same vertical difference ($\delta d = 0.195$ for MM and $\delta d = 0.198$ for Stine) for a value of the experimental density $d = 1.5\ \text{g/cm}^3$.

Moreover two general equations can be derived from the correlation Eqs. 45 and 46.

Combining:

$$\text{for liquids} \qquad d_{\text{Stine}} = 0.916 d_{\text{exp}} + 0.328 \qquad (45)$$
$$\text{and} \qquad d_{\text{Stine.cor}} = d_{\text{exp}}$$

$$\text{for solids} \qquad d_{\text{Stine}} = 0.925 d_{\text{exp}} + 0.116 \qquad (46)$$
$$\text{and} \qquad d_{\text{Stine.cor}} = d_{\text{exp}}$$

gives corrected values of density ($d_{\text{Stine.cor}}$) from the values of density d_{Stine} calculated by the Stine's method:

$$\text{for liquids} \qquad d_{\text{Stine.cor}} = 1.091 d_{\text{Stine}} - 0.358 \qquad (47)$$

$$\text{for solids} \qquad d_{\text{Stine.cor}} = 1.081 d_{\text{Stine}} - 0.126 \qquad (48)$$

Equation 47 will allow to calculate precisely the density of a liquid; this is an extension of the Stine procedure which was designed for solids. On the other hand, the prediction of the density of solids was possible according the Stine's method;[42,43] but Eq. 46 shows that the line for solids has a slope of 0.925 which cuts the ideal line (the slope unity) at a value of $d_{\text{exp}} = 1.546$; in other words, the density values calculated by Stine method are overestimated or underestimated if they are calculated below or above this experimental value $d_{\text{exp}} = 1.546$. This is unexpected. One should expect individual errors in calculations by the Stine's method because the calculated crystal molecular volume (V_{cm}) does not take into account the actual conformational geometries of molecules, but within a large series, this should give random, not systematic errors. Comparison of results with Molecular Mechanics or with the Stine parameters do not agree with this assumption and they show a systematic deviation. A possible explanation could rest in the actual nature of the crystal structure (or packing coefficient). As discussed in Chapter III, the point is that when going from molecular volumes to crystal volumes, empirical parameters usually ignore the packing coefficient or take it as statistically constant. Cady however tried to mimic this packing coefficient by introducing a corrective term in his density calculations:

$$d_{\text{Cady}} = (\text{TDM})k \qquad (49)$$

Besides the theoretical maximum density (TMD explained in Chapter III), the parameter k is expressed as:

$$k = 0.7686 - 0.1280\left(\frac{\text{nb CH}}{TA}\right) \qquad (50)$$

(where nb CH is the number of hydrogen atoms bonded to carbon atoms, and *TA* is the total number of atoms in the molecule). This parameter *k* is supposed to be equivalent to a packing coefficient. Therefore the Cady's method was applied to the calculation of density of the 142 energetic materials just described and discussed above. However the data plotted in Figure 6 show here again (as in Figures 2, 4 and 5) one line for liquids and another line for solids.

The corresponding correlation equations are:

$$\text{for liquids} \qquad d_{\mathrm{Cady}} = 0.992 d_{\mathrm{exp}} + 0.196 \qquad (51)$$

$$\text{for solids} \qquad d_{\mathrm{Cady}} = 0.871 d_{\mathrm{exp}} + 0.163 \qquad (52)$$

Clearly the parameter *k* devised by Cady for incorporating the packing coefficient makes no difference with other empirical or MM approaches. On the other hand combining Eqs. 51 and 52 with $d_{\mathrm{Cady.cor}} = d_{\mathrm{exp}}$ affords the equations:

$$\text{for liquids} \qquad d_{\mathrm{Cady.cor}} = 1.008 d_{Cady} - 0.197 \qquad (53)$$

$$\text{for solids} \qquad d_{\mathrm{Cady.cor}} = 1.148 d_{Cady} - 0.187 \qquad (54)$$

which allow calculations of the density of liquids and solids. Here again,

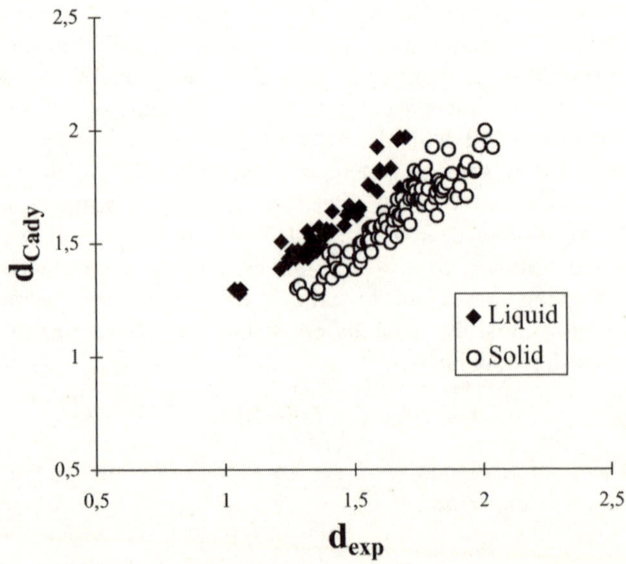

Figure 6. Densities according to Cady (d_{Cady}) versus experimental densities.

correlation analysis in complement to empirical parameters gives a precise and extended (to liquids and solids) methodology for the prediction of density.

In summary the estimation of density by Molecular Mechanics seems presently a good choice. It is easy to operate (it works on a microcomputer); the parameterization is very large (it was devised for structure and reactivity in molecular chemistry) and the precision is equivalent or superior to other methods. Most important it takes into account the actual stereochemical geometry of the molecules.

V. QUANTITATIVE STRUCTURE-PROPERTY RELATIONSHIPS

Quantitative Structure-Property Relationships (QSPR) has been designed primarily to find out correlations between an observed property and the structure of a molecule, especially when there is no clear evidence of which factor(s) of structure is(are) preeminent.

Basic principle involves several steps:

- selection of a set of molecules with known values of the selected property.
- input of these molecules and values in a table with the use of a program able to describe more or less precisely the molecules (e.g. connectivity between atoms, 3D structures, stereochemistry, ...).
- computing descriptors of molecules. These may be topological descriptors (as Wiener, Randic indexes), electronic descriptors (dipole moment in one or several directions, charge density, molecular orbitals, ...), structural descriptors (number of selected types of atoms or functional groups, ...).
- if necessary, adding numerical experimental data supposed to be connected to the property to be predicted and then considered as added potential descriptors.
- performing statistical calculations of all those inputs to highlight correlations and reject irrelevant descriptors by cross-validation. Commercial packages now afford sophisticated mathematical tools like genetic function algorithm.
- predicting desired property of new molecules not included in the learning set.

QSPR is widely used for years in many domains like drug design. Number of applications increases continuously owing to availability of softwares and powerful low cost computers.

Density prediction by QSPR remains less common and only few papers describe efforts in this domain.

Early work seems to have been done by Russian researchers in the field of high explosives by Pivina and coworkers.[83,84]

Their descriptors are fragments of the structural formula. The difference with the above mentioned additive methods lies in the automated computerized generation of all possible fragments. Calculations are then performed by boolean recognition of a fragment in the molecule.

Density is not entered as a figure but molecules are partitioned in four ranges: low, medium, high and ultra-dense ones.

Authors report respectively 92% and 94% of correct classification of test molecules in the low and ultra-dense classes.

Although published results are focused on furazan and furoxan derivatives bearing energetic substituents, the method appears powerful to identify fragments responsible for high density, namely those belonging to Class 4 and not to Class 1 compounds.

Application is claimed for design of new dense explosives by coupling this procedure to an automated generation of structures from an empirical formula, rejecting non-dense molecules, and adding a last step of molecular mechanics or quantum mechanics calculation to predict crystals structures.

Korean workers[85] recently published QSPR prediction of density on the test set of energetic compounds proposed by Piacenza et al.[3] as a basis for molecular mechanics density evaluation.

They used commercial package based on molecular holograms (HQSAR).

For the moment, their results appear to be slightly inferior to MM density determination (Table 19) although authors limited their trials only to solids and discarded some experimental values, and not so general as MM which proved to work either on solids or liquids.

Method performs roughly as well as empirical Cady's[65] or Stine's.[42,43]

Table 19. Comparison of HQSAR with empirical and MM density predictions

Method of calculation	*Number of data*	r^2	*Average error*	*Ref.*
Cady	93 (solids)	0.7268	4.54%	91
Stine	93 (solids)	0.9198	2.31%	91
HQSAR 1	93 (solids)	0.8597	3.01%	91
HQSAR 2	89 (solids)	0.8778	2.66%	91
Cady	142 (solids and liquids)	—	3.02%	3
Stine	142 (solids and liquids)	—	2.95%	3
MM	142 (solids and liquids)	—	2.87%	3

As a conclusion, QSPR methods have been evaluated only in the field of highly energetic molecules. The relatively low number of available data is a limiting factor for extensive calculations. It can be overcome, in next years, by examination of larger sets of data for enlarged applications.

First examples and availability of software will undoubtedly encourage further work in this area of interest. It can be outlined that QSPR presents a major advantage if several properties are to be predicted because only one calculation of descriptors is required.

VI. VARIATION OF DENSITY WITH TEMPERATURE

This paragraph will discuss two different aspects of the problem of variation of density with temperature: the first aspect is related to the prediction of the change of density with temperature of a compound (liquid or solid) for which an experimental value of the density is known. This is a classical problem for chemical engineers; appropriate equations and reference textbooks are known for this purpose. Therefore only a short summary will be given on this topic (vide infra). The second aspect addresses a major question regarding any new compound not yet synthesized. Empirical (or more theoretical) equations have been given and discussed in Paragraphs 2–5 for predicting the density of liquids and solids; however the fundamental question is: will this new compound be a solid or a liquid and therefore what parameters will be used for the prediction: those for solids or those for liquids? The question can be addressed differently, in a way that could give a workable solution: what will be the melting point of the new compound? Given the answer, and depending whether the melting point is above or below ambient temperature, parameters for solids or for liquids will be used. We will discuss this problem in the second part of the paragraph.

Let us first consider the variation of density with temperature of known compounds for which experimental values are available. Generally, this variation is more important for liquids than for solids.[10] This is due probably to weaker intermolecular forces in solution compared to the more rigid structure of crystal lattice. For most pure compound usually at least one experimental saturated liquid density is available.[86] Then the Rackett equation,[87] as modified by Spencer and Danner[88] can be used to calculate saturated liquid densities at absolute temperature T:

$$\rho_{\text{sat}} = P_c/RT_c\ Z_{\text{RA}}^{[1+(1-T_1)2/7]} \qquad (55)$$

where ρ_{sat} = saturated-liquid molar density
P_c = critical pressure
R = gas constant

T_c = absolute critical temperature
Z_{RA} = constant determined from experimental data
T_r = reduced temperature = T/T_c

Errors are less than 1% except for alcohol and acids for which the errors are less than 2%.[89]

Secondly, the general question of predicting the nature of any new compound (will it be a solid or a liquid at ambient temperature?) is related to the prediction of melting point. An excellent recent review; to which readers could refer for details and a thorough discussion, has been published by Dearden[11] in Volume 2 of the series Advances in Quantitative Structure Properties Relationships.

Early studies reported the variation of melting point T_m of hydrocarbons with molecular weight[90–92] or with chain length[93–98] in homologous series. Later studies used topological indices, which gave in some cases fair[99] to good[100] correlation or in other cases were unable to predict the melting point.[101,102] The neural network approach gave promising results.[103] For organic compounds, other than hydrocarbons, several approaches have been developed utilizing physicochemical and structural parameters including: pairs of series;[104] QSAR parameters;[105–107] undefined descriptors (with the CODESSA software);[108] connectivity indices;[102,109] intermolecular force equation;[110] enthalpy of sublimation and of vaporization[111] or solubility.[112–115] Some of these methods gave satisfactory results; others seem of marginal interest. It should be noted that any of these methods, which uses a physico-chemical parameter, is useless for a new compound to be made since none of these physico-chemical parameter is available in this case.

The group contribution approach is among the most recent methodologies. Since the melting point is controlled mostly by the enthalpy of fusion, and to a lesser extent by the entropy of fusion, a simple approach ignores the entropy of fusion.[116,117]

A more complete approach incorporates an entropy contribution.[118–121] Predictions using group contribution methods seem good to very good; among these studies the Yalkowski's group contribution[118–121] appears to Dearden[11] to be the simplest of the present available methods.

It is likely that the most workable and precise calculation procedures use equations which have been introduced in computerized methods. Unfortunately some of these equations need physical constants which are not available for compounds which have not yet been prepared. This is the case of the equation of Lorenz and Herz[122] (Eq. 56)

$$\frac{T_m}{T_b} = 0.5839 \qquad (56)$$

which is proposed in the programs CHEMEST,[123] QSAR[124] and

MPBPVP.[125] Equation 56 contains a boiling point (T_b) whose value is not available (for a compound to be prepared). One possibility would be to displace further the question and to calculate the boiling point by the method of Stein and Brown[126] (this is the case in the MPBPVP program[125]). However this procedure increases the error in the estimation of T_m.

The program CHEMEST also contains the method of Grain and Lyman[127] (Eq. 57)

$$T_m = 0.474\,(10^{3-4/\sqrt{M}}) + b\rho_b - 28 \tag{57}$$

which is useless for further predicting the density because Eq. 57 needs the value of the liquid density at normal boiling point (a case of vicious circle). On the other hand, the method of Joback[116,117] calculates the melting point with group increments (Eq. 58)

$$T_m = 122.5 + \Sigma GC \tag{58}$$

and is available in the programs MPBPVP[125] and Chem Prop Pro.[128]

VII. CONCLUSIONS

As conclusions to this chapter, general comments can be made.

Concerning the evolution of predictive methods, for the estimation of density, there is a close similarity with other physico-chemical parameters. The fact that quite acceptable results have been obtained as early as the beginning of the 19th century, is still surprising referring to the basic chemical knowledge at that time. The work of Exner deserves a special comment because it was really a pioneering contribution to the field of group increments procedures. With time, empirical methods have gained in generality and in precision. They are easy to use and they give good results, but referring to stereochemistry, they remain two dimensional procedures.

One may have expected that more theoretical methods, that take into account the actual three dimensional structure of the molecule, would replace completely previous empirical procedures and correlation analysis. This is not completely true. Not only because empirical parameters are still valid, since they give reasonably accurate results without sophisticated computation, but also because theoretical calculations can combine with (and not necessarily compete with) correlation analysis. This is the case with Molecular Mechanics. Taken independently, Molecular Mechanics would have given only crude and limited estimations of density; it is only the combination of Molecular Mechanics and of correlation analysis that gave precise and general (for solids and liquids) estimation of density. In addition the use of correlation analysis has been extended to group increment parameters giving more precise values of density for both solids and liquids.

Referring to a near future, important progress can be anticipated, during the years to come, from two methodologies: theoretical calculations directed towards crystal packing, and quantitative structure properties relationships.

ACKNOWLEDGMENT

The authors wish to acknowledge the contribution of co-workers at SNPE and at ENSSPICAM for the development of this topic. They are grateful to Dr B. Blaive (Aix-Marseille 3 University) for access to, and helpful suggestions in the use of, EMO and satellite programs. The financial support of DGA for projects related to this topic is gratefully acknowledged.

REFERENCES

1. Sax, N. I.; Lewis, R. J. *Hawley's Condensed Chemical Dictionary*; 13th ed.; Van Nostrand Ranhold: New York, **1997**.
2. Altmann, K. L.; Chafin, A. P.; Merwin, L. H.; Wilson, W. S.; Gilardi, R. *J. Org. Chem.* **1998**, *63*, 3352–3356.
3. Piacenza, G.; Jacob, G.; Graindorge, H.; Blaive, B.; Gallo, R. Combustion and Detonation ICT Annual Conference, Karlsruhe, **1997**, pp. 123-1–123-14.
4. Subramanian, G.; Eck, G.; Boyer, J. H.; Stevens, E. D.; Trudell, M. L. *J. Org. Chem.* **1996**, *61*, 5801.
5. Marchand, A. P.; Zope, A.; Zaragoza, F.; Bott, S. G.; Ammon, H. L.; Du, Z. *Tetrahedron* **1994**, *50*, 1687.
6. Krevelen, D. W. v.; Chermin, H. A. G. *Fuel* **1954**, *33*, 79.
7. Elbro, H. S.; Fredenslund, A.; Rasmussen, P. *Ind. Eng. Chem. Res.* **1991**, *30*, 2576–2582.
8. Krevelen, D. W. v.; Hoftyzer, P. V. *J. Appl. Polym. Sci.* **1969**, *13*, 871–881.
9. Lide, D. C.; Kehiaian, H. V. *CRC Handbook of Thermophysical and Thermodynamic Data*; CRC Press: Boca Raton, **1994**.
10. *Kirk–Othmer Encyclopedia of Chemical Technology*; 4th ed.; Wiley, J., Ed.: New York, **1992**, *4*.
11. (a) Dearden, J. C. In *Advances in Quantitative Structure Properties Relationships*; JAI Press: Stamford, **1999**, Vol. 2, pp. 127–175. (b) Cohen, N.; Benson, S. W. *Chem. Rev.* **1993**, *93*, 2419–2438.
12. Kopp, H. *Ann. Phys.* **1839**, *47*, 133–153.
13. (a) Traube, J. *Ber.* **1895**, *28*, 2722. (b) Traube, J. *Ann. Phys.* **1907**, *22*, 519.
14. (a) Le Bas, G. *Chem. News.* **1911**, *102*, 226. (b) Le Bas, G. *Chem. News.* **1913**, *104*, 199.
15. Davis, H. G.; Gottlieb, S. *Fuel* **1963**, *42*, 37.
16. Harrison, E. K. *Fuel* **1965**, *44*, 339.
17. (a) Girolami, J. *J. Chem. Ed.* **1994**, *71*, 964. (b) Gordon, J. A.; Ford, R. A. *The Chemist Companion*, Wiley: New York, **1972**.
18. Bondi, A. *J. Phys. Chem.* **1964**, *68*, 441.
19. Exner, O. *Correlation Analysis of Chemical Data*; Plenum Press: New York, **1988**.
20. Shorter, J. *Correlation Analysis of Organic Reactivity*; RSP: Chichester, **1982**.

21. Unger, S. H.; Hansch, C. *Prog. Phys. Org. Chem.* **1976**, *72*, 91.
22. Charton, M. In *Classical and 3-D QSAR in Agrochemistry and Toxicology*; ACS, Ed. Washington D. C., **1995**.
23. Exner, O. *Collection Czech. Chem. Commun.* **1966**, *31*, 3222–3251.
24. Sugden, S. *J. Chem. Soc.* **1924**, *125*, 1177–1189.
25. Quayle, O. R. *Chem. Rev.* **1953**, *53*, 439.
26. Vogel, A. I.; et al. *J. Chem. Soc.* **1961**, 570.
27. Ahmad, P.; Fyke, C. A.; Mellors, A. *Biochem. Pharmacol.* **1975**, *24*, 1103–1109.
28. Exner, O. *Nature* **1962**, *195*, 890–891.
29. Exner, O. *Collection Czech. Chem. Commun.* **1967**, *32*, 1–23.
30. Nielsen, A. T. Naval Weapons Center Report, **1973**, NWC TP 5452.
31. Tarver, C. M. *J. Chem. Eng. Data* **1979**, *24*, 136–145.
32. Benson, S. W.; Bush, J. H. *J. Chem. Phys.* **1958**, *29*, 546.
33. Benson, S. W.; Cruickshank, F. R.; Golden, D. M.; Haugen, G. R.; O'Neal, H. E.; Rodgers, A. S.; Shaw, R.; Walsh, R. *Chem. Rev.* **1969**, *69*, 279–324.
34. Note that Benson has made corrections of nearest neighbour interactions.
35. Eremenko, L. T. *Izv. Akad. Nauk. SSSR, Ser. Khim. (Engl. Transl.)* **1968**, 1012–1016.
36. Coulson, C. A. *Valence*; 2nd Ed.; Oxford University Press: Oxford, **1961**.
37. Eremenko, L. T. *Izv. Akad. Nauk. SSSR, Ser. Khim. (Engl. Transl.)* **1970**, 46–52.
38. Eremenko, L. T. *Izv. Akad. Nauk. SSSR, Ser. Khim. (Engl. Transl.)* **1972**, 172–174.
39. Eremenko, L. T.; Korolev, A. M. *Izv. Akad. Nauk. SSSR, Ser. Khim. (Engl. Transl.)* **1972**, 2742–2744.
40. Eremenko, L. T. Proc. Int. Pyrotech. Semin., Arcachon, France, **1982**, pp. 740–758.
41. Eremenko, L. T. Proc. 11th Symp. Expl. Pyrotech., Philadelphia (PA), **1981**, pp. 29/1–29/9.
42. Stine, J. R. Los Alamos National Laboratory Report, **1981**, LA-8920.
43. (a) Stine, J. R. In *Structure and Properties of Energetic Materials*, Vol. 296, MRS: Pittsburgh, 1993. (b) Ammon, H. L.; Mitchell, S. *Prop. Exp. Pyrotech.* **1998**, *23*, 260–265.
44. Kitagorodski *Molecular Crystals and molecules*; Academic Press: New York, **1973**.
45. Gavezotti, A.; Filippini, G. *Acta Cryst.* **1992**, *B48*, 537–545.
46. Gavezotti, A.; Filippini, G. *J. Phys. Chem.* **1994**, *98*, 4831–4837.
47. (a) Wolff, J. J. *Angew. Chem. Int. Ed. Engl.* **1996**, *35*, 2195–2197. (b) Gdanitz, R. J. In *Theoretical Aspects and Computer Modeling of the Molecular Solid State*, Gavezotti, A. Ed, Wiley: Chichester, **1997**.
48. Williams, D. E. *Science* **1968**, *159*, 645–646.
49. Williams, D. E. *Acta Cryst.* **1969**, *A25*, 464–470.
50. Williams, D. E. *Acta Cryst.* **1972**, *A28*, 629–635.
51. Williams, D. E. *Acta Cryst.* **1973**, *A29*, 409–412.
52. Williams, D. E.; Starr, T. L. *Comput. Chem.* **1977**, *1*, 173–177.
53. Dunitz, J. D.; Gavezotti, A. *Acc. Chem. Res.* **1999**, *32*, 677–684.
54. Gdanitz, R. J. *Chem. Phys. Lett.* **1992**, *190*, 391–396.
55. (a) Holden, J. R.; Du, Z.; Ammon, J. L. *Comput. Chem.* **1993**, *14*, 422–437; (b) Dzyabchenko, A. V.; Pivina, T. S.; Arnautova, E. A. *J. Mol. Struct.* **1996**, *378(2)*, 67–82.
56. Karfunkel, H. R.; Gdanitz, R. J. *J. Comput. Chem.* **1992**, *13*, 1171–1183.
57. Eijck, B. P. v.; Modij, W. T.; Kroon, J. *Acta. Cryst.* **1995**, *B51*, 99–103.
58. Williams, D. E. *Acta Cryst.* **1996**, *A52*, 326–328.
59. Gavezotti, A. *J. Am. Chem. Soc.* **1991**, *113*, 4622–4629.
60. Gavezotti, A. *Acc. Chem. Res.* **1994**, *27*, 309–314.
61. Gavezotti, A.; Filippini, G. *J. Am. Chem. Soc.* **1995**, *117*, 12299–12305.
62. Gavezotti, A. *Acta Cryst.* **1996**, *B52*, 201–208.
63. Gdanitz, R. J.; Leusen, F. J. J.; Karfunkel, H. R. *J. Mol. Graphics.* **1993**, *11*, 275.

64. Immirzi, A.; Perini, B. *Acta Cryst.* **1977**, *A33*, 216–218.
65. (a) Cady, H. H. Los Alamos National Laboratory Report, LA-7760-MS, **1979**. (b) Cromer, D. T.; Ammon, H. L.; Holden, J. R. Los Alamos National Laboratory Report, LA-11142-MS, **1987**.
66. (a) Allinger, N. L. Program QCPE-395, Department of Chemistry, Indiana University, Bloomington, IN-47405. (b) Burkert, U.; Allinger, N. L. *Molecular mechanics*, ACS Monograph 177, ACS Ed.: Washington D.C., **1982**.
67. Osawa, E.; Musso, H. *Angew. Chem., Int. Ed. Engl.* **1983**, *22*, 1.
68. Special issue on Molecular Mechanics and Modeling, *Chem. Rev.* November **1993**, *93*, n°7.
69. Rappe, A. K.; Casewitt, C. J. *Molecular Mechanics across Chemistry*; University Science Books: Sansalito (CA), **1997**.
70. Lipkowitz, K. B.; Peterson, M. A. *Chem. Rev.* **1993**, *93*, 2463–2486.
71. Eksterowicz, J. E.; Houk, K. N. *Chem. Rev.* **1993**, *93*, 2439–2462.
72. Rappe, A. K.; Casewitt, C. J. *Molecular Mechanics across Chemistry*; University Science Books: Sansalito (CA), **1997**.
73. Landis, C. R.; Root, D. M.; Cleveland, T. *Reviews in Computational Chemistry*; VCH: New York, **1995**.
74. Galiatsatos, V. *Reviews in Computational Chemistry*; VCH: New York, **1995**.
75. Rappe, A. K.; Casewitt, C. J. *Molecular Mechanics across Chemistry*; University Science Books: Sansalito (CA), **1997**.
76. Doucet, J. P.; Weber, J. *Computer aided Molecular Design*; Academic Press: London, **1996**.
77. Conelly, M. L. *Science* **1981**, *211*, 661–666.
78. Stouch, T. R.; Jurs, P. C. *J. Chem. Info. Comput. Sci.* **1986**, *26*, 4–12.
79. Higo, J.; Go, N. *J. Comput. Chem.* **1989**, *10*, 376–379.
80. Gavezotti, A. *J. Am. Chem. Soc.* **1983**, *105*, 5220–5225.
81. Piacenza, G.; Legsaï, G.; Blaive, B.; Gallo, R. *J. Phys. Org. Chem.* **1996**, *9*, 427–432.
82. The method for calculating the MM molecular volume (V_{molMM}), computed by B. Blaive, was first disclosed in: Delpeyroux, D.; Blaive, B.; Gallo, R.; Becuwe, A.; Piteau, M.; Jacob, G.; Gautier, J-C. Energetic Materials ICT Annual Conference, Karlsruhe **1993**, pp. 74–1; and fully described in Blaive, B.; Legsaï, G.; Boileau, J. Europyro, Vaires/Marne, **1995**.
83. Pivina, T. S.; Schlyapochnikov, V. A. *Zh. Fiz. Khim.* **1992**, *66*, 84–100.
84. Pivina, T. S.; Schlyapochnikov, V. A. Proc. Int. Pyrotech. Semin., Beijing, **1991**, pp. 212–219.
85. Cho, S. G.; Park, B. S. NDIA Conference, Tampa (Florida), **1999**, pp. 455–465.
86. Reid, R. C.; Pransnitz, J. M.; Sherwood, T. K. *The Properties of Gases and Liquids*; 3rd Ed.; Mac-Graw Hill: New York, **1977**.
87. Rackett, H. G. *J. Chem. Eng. Data* **1970**, *15*, 514.
88. Spencer, C. F.; Danner, R. P. *J. Chem. Eng. Data* **1972**, *17*, 236.
89. Spencer, C. F.; Adler, S. B. *J. Chem. Eng. Data* **1978**, *23*, 82.
90. Austin, J. B. *J. Am. Chem. Soc.* **1930**, *52*, 1049–1053.
91. Etessam, A. H.; Sawyer, M. F. *J. Int. Petrol* **1939**, *25*, 253–262.
92. Keyes, R. W. *Phys. Rev.* **1959**, *115*, 564–567.
93. Garner, W. E.; Bibber, K. v.; King, A. E. *J. Chem. Soc.* **1931**, 1533–1541.
94. Merkel, J. H. *Proc. Royal Acad. Amsterdam* **1937**, *40*, 164–173.
95. Meyer, K. H.; Wyk, A. v. d. *Helv. Chim. Acta* **1937**, *20*, 1313–1320.
96. Gray, C. J. *J. Inst. Petrol* **1943**, *29*, 226–234.
97. Broadhurst, J. *J. Res. Nat. Bur. Sts* **1962**, *66A*, 241–249.
98. Marano, J. J.; Holder, G. D. *Ind. Engin. Chem. Res.* **1997**, *36*, 1895–1907.
99. Seybold, P. G.; May, M. A.; Gargas, M. L. *Acta Pharm. Yugosl.* **1986**, *36*, 253–265.
100. Hanson, M. P.; Rouvray, D. H. In *Graph theory and Topology in Chemistry*; King, R. B.; Rouvray, D. H., Eds; Elsevier: Amsterdam, **1987**, pp. 201–208.

101. Needham, D. E.; Wei, I. C.; Seybold, P. G. *J. Am. Chem. Soc.* **1988**, *110*, 4186–4194.
102. Pogliani, L. *J. Phys. Chem.* **1995**, *99*, 925–937.
103. Cherqaoui, D.; Villemin, D.; Kvasnicka, V. *Chemom. Intell. Lab. Systems* **1994**, *24*, 117–128.
104. *QSAR in Design of Bioactive Compounds*; Charton, M.; Charton, B. I., Eds.; J. R. Prous: Barcelona, **1984**.
105. Dearden, J. C.; Rahman, M. H. *Math. Comput. Modelling* **1988**, *11*, 843–846.
106. Verloop, A.; Hooganstraaten, W.; Tripker, J. *Drug Design*; Academic press: New York, **1976**.
107. Dearden, J. C. *Sci. Total. Environ.* **1991**, *100/110*, 59–68.
108. Murugan, R.; Grendze, M. P.; Toomey, J. E.; Katritzky, A. R.; Karelson, M.; Lobanov, V.; Rachwal, P. *CHEMTECH* **1994**, *24*, 17–23.
109. Medic-Savic, M.; Nickelic, S.; Matijevic-Sosa, J. *Acta Pharm.* **1992**, *42*, 153–167.
110. Charton, M.; Charton, B. I. *J. Phys. Org. Chem.* **1994**, *7*, 196–206.
111. Westwell, M. S.; Searle, M. S.; Wales, D. J.; Williams, D. H. *J. Am. Chem. Soc.* **1995**, *117*, 5013–5015.
112. Yalkowsky, S. H.; Valvani, S. C.; Roseman, T. J. *J. Pharm. Sci.* **1983**, *72*, 866–870.
113. Anderson, B. D.; Conradi, R. A. *J. Pharm. Sci.* **1985**, *74*, 815–820.
114. Rubino, J. T. *J. Pharm. Sci.* **1989**, *78*, 485–489.
115. Thomas, E.; Rubino, J. T. *Int. J. Pharm.* **1996**, *130*, 179–183.
116. Joback, K. C.; Reid, R. C. *Chem. Engin. Comm.* **1987**, *57*, 233–243.
117. *The Properties of Gases and Liquids*; 4th Ed.; Reid, R. C.; Pransnitz, J. M.; Poling, B. E., Eds.; McGraw-Hill: New York, **1987**.
118. Simamora, P.; Yalkowsky, S. H. *Environm. Res.* **1993**, *1*, 293–300.
119. Simamora, P.; Miller, A. H.; Yalkowsky, S. H. *J. Chem. Inf. Comput. Sci.* **1993**, *33*, 437–440.
120. Simamora, P.; Yalkowsky, S. H. *Ind. Eng. Chem. Res.* **1994**, *33*, 1405–1409.
121. Krzyzaniak, J. F.; Myrdal, P. B.; Simamora, P.; Yalkowsky, S. H. *Ind. Eng. Chem. Res.* **1995**, *34*, 2530–2535.
122. Lorenz, R.; Herz, W. *Anorg. Allgem. Chem.* **1922**, *122*, 51–60.
123. Lyman, W. J.; Potts, R. G.; Magil, G. C. *User's Guide to CHEMEST*: Cambridge (MA), **1984**.
124. Hunter, R.; Faulkner, L.; Culver, F.; Hill, J. *QSAR Structure-Activity based Chemical Modeling and information Software*; Montana State University: Montana, **1985**.
125. Syracuse Research Corporation MPBPVP PC-based program ver. 1.25; Syracuse, NY, 1997.
126. Stein, S. E.; Brown, R. L. *J. Chem. Info. Comput. Sci.* **1994**, *34*, 581–587.
127. Grain, C. F.; Lyman, W. J. Interim Report on Task 29 of Environmental Protection Agency, U.S. Environmental Protection Agency Office of Toxic substances, Washington DC, **1983**.
128. Cambridge Soft Corporation, Chem Prop Pro PC-based program; Cambridge (MA), **1998**.

HALOGENS IN MOLECULAR STRUCTURE–PROPERTY RELATIONSHIPS

Paul G. Seybold

Advances in Quantitative Structure Property Relationships
Volume 3, pages 109–136.

I. INTRODUCTION

Halogens are common substituents on organic compounds and can profoundly alter the properties of these compounds. In large part because of this ability, halogenated compounds have found wide use commercially as refrigerants, degreasers, cleaning agents, solvents, flame retardants, electrical insulators, propellants, pesticides, herbicides, anesthetics, drugs, and in other roles. More recently the roles of halogenated compounds in ground, water, and air pollution and in toxicology have become issues of national and international concern. In addition, the production of halogenated compounds by natural means, both biological and nonbiological, has become increasingly recognized.

Within the quantitative structure-activity relationship (QSAR) and quantitative structure-property relationship (QSPR) paradigms the halogens form an especially appealing and informative series. Halogens generally appear as single-atom substituents (and only occasionally as parts of groups such as CF_3), and the key properties of these elements, such as their electronegativities, polarizabilities, ionization potentials, etc. vary uniformly in the order of their atomic numbers. As a result, the variation in a property that takes place as a compound is successively substituted with fluorine, chlorine, bromine, and iodine substituents can be highly informative regarding the underlying forces or interactions involved in the compound's activity.

This review focuses on how halogens can be treated in traditional QSAR and QSPR studies. Other interesting approaches, such as neural networks, COMFA,[1] and the partial least squares (PLS) approach,[2–4] despite their usefulness, lie outside the present purview. First, the electronic natures of the halogens will be examined. Following this preliminary overview, the various means for treating halogens within the multiple linear regression (MLR) approach will be described, starting with the most simple approaches and progressing to more advanced techniques. Many of the methods described will be seen to be generally applicable to other heteroatoms as well. Because of the great number of papers that have treated halogenated compounds, either selectively or as parts of larger diverse data sets, no attempt has been made to give a complete review of this vast literature; rather, the aim of this review is to inform readers of the major approaches available for the treatment of halogens in QSAR and QSPR.

II. THE NATURE OF HALOGENS

The central problem facing structure-property applications involving halogens is that these elements have split personalities, and show apparently anomalous behavior in a number of contexts. (The term "anomalous" is used because we don't fully understand it.) In organic chemistry, halogens are known to be *ortho*/*para*-directing agents.[5] For example, the nitration of fluorobenzene is found to lead to 13% *ortho*-nitration, 86% *para*-nitration, and only 1% *meta*-nitration. Nitration of chlorobenzene occurs 35% at the *ortho* position, 64% at the *para* position, and 1% at the *meta* position. Bromo- and iodobenzene show similar trends.[5] Thus, with respect to their interactions with the pi-electron systems of conjugated and aromatic compounds, these elements might be considered electron donors. However, it has long been recognized that in many instances the halobenzenes are *less* reactive toward electrophilic attack than is benzene itself, and the situation, especially in the case of fluorine, is clearly more complex than simple electron donation.[6,7]

On the other hand, the halogen atoms F, Cl, and Br are more electronegative (χ) than carbon, and therefore these atoms function as electron-withdrawing agents within the sigma electronic systems of compounds to which they are attached. Fluorine is the most electronegative element in the periodic table. Chlorine and bromine are less electronegative than fluorine, but more electronegative than carbon, and iodine has an electronegativity comparable to carbon's. In addition, the atomic *polarizabilities*—representing the ease with which the electronic clouds of these elements are distorted by an imposed electric field—vary in the opposite order to the atomic electronegativities. Applequist and coworkers have determined a number of atom polarizabilities using an atom dipole interaction model.[8–10] The atom polarizability is an important factor in determining the strength of van der Waals intermolecular forces, in particular the induction and dispersion forces, which are responsible for attractive interactions between molecules. Unless the molecules are very highly polar, dispersion forces are the usually the dominant forces governing these interactions. The atomic electronegativities and polarizabilities of the key elements involved are shown in Table 1.

Are all fluorines (or all chlorines, bromines, or iodines) in molecules alike? That is, can we expect that once we understand the nature (e.g. charge, polarizability, etc.) and role of a halogen in one compound that this understanding can be transferred to the same halogen appearing in other compounds? Most likely not, since the physical, chemical, and biological influences of any atom or group will depend on its partial electronic charge Q, and this charge in turn will depend on both the nature of the atom itself

Table 1. Electronegativities, first ionization potentials, and atomic polarizabilities of carbon and the halogens

	Carbon	*Hydrogen*	*Fluorine*	*Chlorine*	*Bromine*	*Iodine*
Electronegativity[a]	2.55	2.20	3.98	3.16	2.96	2.66
Ionization potential[b]	1086.2	1312.0	1681	1251.1	1139.9	1008.4
Atomic refraction[c]	2.418[d]	1.100	0.95	5.967	8.865	13.900
Atom polarizability[e] (Å^3)	0.777	0.172	0.352	1.761	2.819	4.69[f]

Notes: [a] Pauling electronegativity.
[b] Units: kJ/mol.
[c] *Physical Properties of Chemical Compounds—II*, Advances in Chemistry Series, No. 22, American Chemical Society, Washington, D.C., 1959, p. 8.
[d] Singly bound.
[e] Ref. 10.
[f] Ref. 8.

and the natures of the other atoms in the molecule that are competing for the molecule's electrons. The electronegativity χ of an atom depends on its charge Q, as described most simply by a Taylor series expansion of χ about $Q=0$,

$$\chi = \chi_0(Q = 0) + (\partial\chi/\partial Q)_{Q=0} + (\partial^2\chi/\partial Q^2)_{Q=0} + \cdots$$

An atom's electronegativity falls as it acquires greater negative charge, and rises if it becomes more positive. The polarizability α of the atom behaves in the opposite manner, increasing as negative charge is acquired and falling as it is lost.[11] As explained by Politzer,[12] the second term on the right is the reciprocal of what Huheey has called the "charge capacity" κ of the atom,[13–16] i.e.

$$\kappa = 1/(\partial\chi/\partial Q)_{Q=0}$$

In Huheey's original definition the series is truncated after the second term on the right, and $\kappa = 1/(IP - EA)$,[12,16] where IP is the ionization energy and EA is the electron affinity of the atom. It can be shown that κ is related to the hardness of the neutral atom.[12] The charge capacity describes the attraction of the atom for electronic charge and its ability to absorb that charge. Fluorine has the highest intrinsic electronegativity χ_0 in the periodic table, but as a small atom it lacks the ability to absorb great amounts of charge without a large change in its electronegativity, i.e. it has a low charge capacity. In contrast, chlorine and bromine, although relatively electronegative compared to many elements, have comparably (to fluorine) large charge capacities. Since the performance of an atom in a molecule is generally strongly related to its electronegativity and polarizability, ultimately the action of a halogen in a molecule will depend on its total molecular environment and how it competes for the available charge in that environment, although in some cases the approximation of atom (or group) transferability may hold to some acceptable degree.

A preliminary look at the variation of halogen charge with surroundings is thus in order, since the key properties of electronegativity and polarizability depend strongly on atomic charge. Unfortunately, the charge on an atom in a molecule is not a properly defined quantum mechanical variable, since the electronic probability distribution of a molecule is spread throughout space and not localized on its atoms. Any definition of atomic charge in a molecule is therefore arbitrary to some extent, and its usefulness for applications such as intended here must be tested. A number of different quantum chemical representations of atomic charges have been presented,[17–20] including the more familiar Mulliken and electrostatic charges, the *natural charges* derived by Weinhold and coworkers from analysis of natural bond orbitals,[21,22] and charges derived from Bader's "atoms in molecules" (AIM) approach.[23,24] In addition to these quantum chemical definitions, several topological schemes, mostly based on Sanderson's concept of electronegativity equilization, have been proposed.[25–30] In some cases these latter schemes have been applied to properties of halocarbons.[29–32]

One reasonable criterion for a useful definition of the charge of an atom in a molecule is that the charge calculated in this manner should vary in a manner consistent with chemical intuition and that the charge variations should correlate sensibly with physically measurable properties.[33] Recently Gross and Seybold[34] have tested several atomic charge definitions for their ability to act as theoretical descriptors for the measured pK_a's of a series of substituted anilines. These authors performed quantum chemical calculations at the Hartree Fock 6–311G** level on these compounds. It was found that the natural charge on the amino nitrogen atom provided the most effective descriptor for this purpose, whereas the Mulliken and electrostatic charges did not correlate with the pK_a's. The natural charges were also effective descriptors for correlating the pK_a changes in a series of substituted phenols.[35] In other cases Bader's AIM charges have also proven to be valuable.

Figure 1 illustrates the variations in the natural charges (Q_{nat}) and AIM charges (Q_{AIM}) calculated at the density functional theory B3LYP/6–311G** level for a series of methanes as the compounds are increasingly substituted with fluorine atoms.[36] Although the two charge measures differ in the absolute values assigned to the atoms, it is evident that the strongly electronegative fluorine atoms largely retain their charge character even in the face of competition from additional fluorine atoms bonded to the same carbon. In contrast, when chlorine atoms are substituted, the less electronegative and more polarizable chlorines are much more sensitive to their molecular environment, to the extent that their charges reverse from negative to positive as more chlorines are attached. Bromines are still more sensitive to charge competition from other atoms.[36] Similar results are found for halogenated ethanes, ethylenes, and benzenes:[36,37] whereas fluorines show relatively little loss of charge as a result of charge competition from additional fluorine atoms attached to the molecular skeleton, chlorines and bromines are highly sensitive both to the presence of other electronegative atoms and to the positions of these atoms on the molecular skeleton.[36]

A second consideration also enters the halogen picture. In an extensive study, Balaban et al. examined the normal boiling points (BPs) of 532 halo- and polyhaloalkanes with 1–4 carbon atoms.[38] They observed several features of interest in the present context. For example, they found that the dependence of the boiling points of the halomethanes on the numbers of

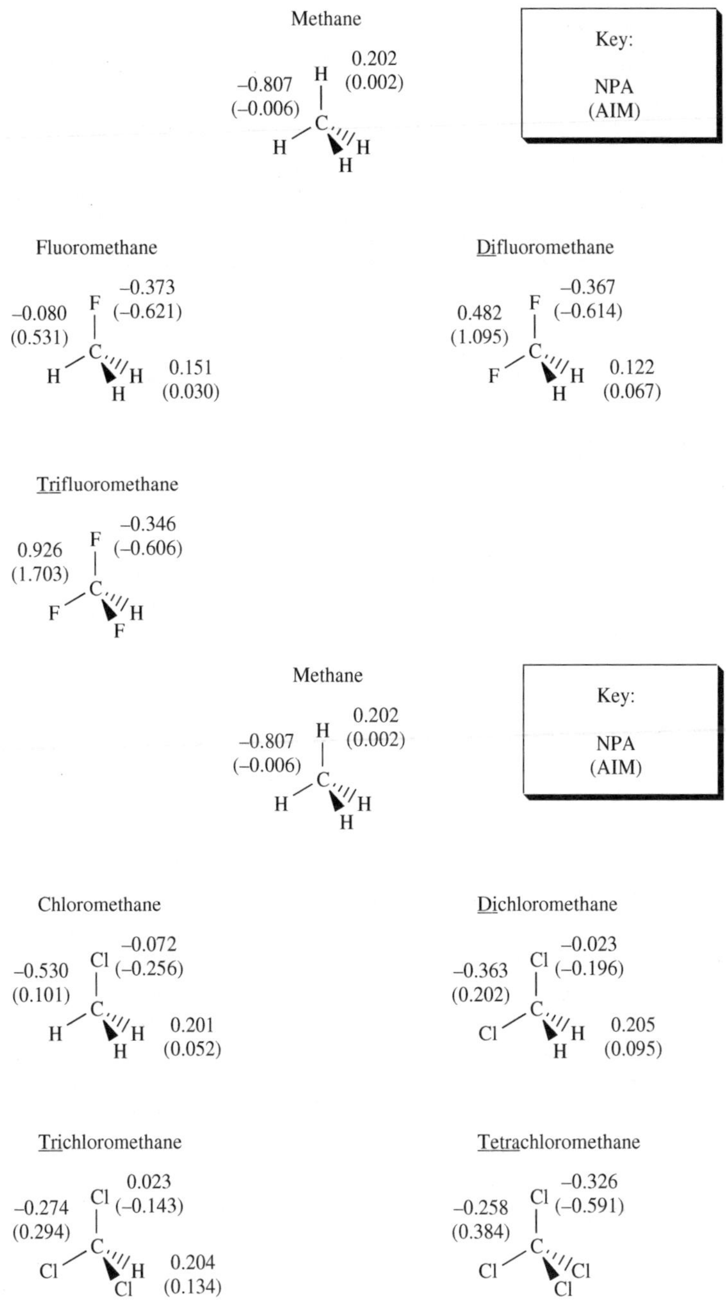

Figure 1. Atomic charges from natural polulation analysis and Bader's AIM method. (From Gross, Seybold, and Hadad, Ref. 36.)

halogens was decidedly nonlinear. For fluoromethanes the boiling points reached a peak at two fluorines, then decreased. Although the authors restricted their attention to noting trends in the experimental data, this observation suggests that specific *intermolecular interactions*, e.g. *pseudo* hydrogen bonds between the positive hydrogens and the negative halogens of these compounds, may strongly influence the boiling points and, by inference, other physical properties. Balaban et al. also made a number of additional observations of interest, e.g. that (1) replacing a geminal F atom on a fluorocarbon by a Cl atom increases the BP by about 45 °C, replacing the F by Br increases the BP by 75°, and replacing the F by I increases the BP by 115°, (2) for dihalo-*n*-alkanes, geminally-substituted compounds boil at lower temperatures than vicinally-substituted ones, and (3) terminally-substituted halocompounds tend to boil at higher temperatures than comparable compounds with interior substitution.[38] These observations also point to effects related to charge- and polarizability-related factors governing the intermolecular interactions.

With the above considerations in mind, it is useful to divide structure-property studies into three categories depending on the type of property considered: *intensive*, *extraactive*, and *intraactive*. By *intensive* one means properties such as the index of refraction or molar refraction that are, for all practical purposes, dependent only on the nature of the isolated single molecule, even though a bulk summation property may be what is actually measured. Such properties do not depend on intermolecular interactions, and often can be reasonably represented in terms of sums of atomic contributions. *Extraactive* properties are those, such as chromatographic retentions and drug-receptor binding constants, which depend on interactions between the molecules in question and some external entity such as a stationary phase or a receptor protein. These properties often depend on specific intermolecular interactions, which in some cases may be geometry dependent, but may in other cases involve nonspecific interactions as well. The actions of some general anesthetics might be included in this category, as when nonpolar molecules disrupt the integrity of cellular membranes.[39] The final category, *intraactive* properties, refers to those properties, such as the boiling point, heat of vaporization, critical temperature, critical pressure, surface tension, second virial coefficient, etc. which depend on interactions between molecules of the same kind in bulk. It is to be expected that different considerations may arise in choosing descriptors for these different categories of properties, although some extent of descriptor overlap is inevitable.

Therefore, to return to the question of whether the properties of the individual halogens can be transferred from one compound to another, it is clear that any approach which fails to account for the complexity of the factors described above is unlikely to be successful. Whereas fluorine atoms largely retain their charge character (and presumably their effective electronegativities, polarizabilities, etc.) in different molecular situations, the charges of the other halogens are much more sensitive to their *intra*molecular bonding environments. Beyond this, *inter*molecular interactions among the compounds, possibly involving polar hydrogen atoms, must also be taken into account for properties, such as the boiling points, heats of vaporization, critical pressures and temperatures, etc. which depend on such interactions. Consequently, one should proceed with care in assessing the behaviors of fluorine, chlorine, bromine, and iodine atoms in different molecular situations.

III. METHODS

A number of different approaches have been used to treat halogens in QSAR and QSPR studies. Some of the most commonly employed of these approaches will now be described.

A. Empirical Approaches

The simplest possible approach to treating halogens in a multiple linear regression (MLR) structure-property analysis is to use as descriptors simple counts of the numbers of the different halogens, as N_F (fluorines), N_{Cl} (chlorines), N_{Br} (bromines), and N_I (iodines). Group contribution approaches of this general nature have been applied frequently, for example in work on the vapor pressures,[40,41] boiling points,[42,43] and other properties of perfluoroalkanes, which are of interest as potential blood replacements.[44] Although this approach can yield good results in some restricted cases, for the reasons given above it often gives only a modest account of the properties considered. In these latter cases chemical intuition can sometimes point to modifications or additions to the simple method that significantly improve the quality of the regression obtained.

An example can be given for the case of the gas chromatographic retentions of a set of polyhalogenated biphenyls. These compounds are of strong interest because they are widespread environmental contaminants with suspected toxic and carcinogenic activities. Höfler et al.[45] measured the

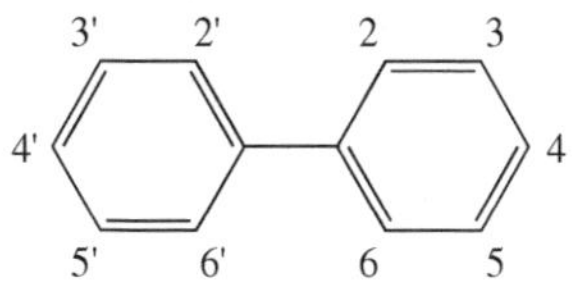

Biphenyl

gas and high-pressure liquid chromatographic retentions of 57 mono- and poly-halogenated biphenyls. They also looked at relationships between the observed retentions and the molecular surface areas of the compounds. Several additional MLR approaches, some of which will be described later, were applied to this data set.[46–48] Although these studies often provided regressions of good quality, some were criticized for the difficulty of calculating the descriptors employed, or on the basis that a physical interpretation of the results was not readily apparent.[48,49] Since it had been recognized from the beginning that halogens at the *ortho* positions were less effective in prolonging GC retention than those at other positions,[45] it seemed reasonable to simply distinguish such halogens from their counterparts in a regression analysis.[50] Indeed this tactic yielded an excellent and readily interpreted structure-property relationship for the retention indices (RI):

$$\mathrm{RI} = 1631 + 231\mathrm{N_{Cl}} + 320\mathrm{N_{Br}} + 440\mathrm{N_{I}} - 169\mathrm{N_{OrCl}} - 175\mathrm{N_{OrBr}} - 214\mathrm{N_{OrI}} \quad (1)$$

$$N = 53 \qquad r^2 = 0.990 \qquad s = 42 \qquad F = 736$$

Here the $\mathrm{N_{Ori}}$ parameters are the corrections for the *ortho* positions. Physically, this was interpreted to imply that halogens at the *ortho* positions were less accessible to the stationary phase.[50] (Previously, the effect of ortho halogens was often attributed to their influence in increasing the nonplanarity of these compounds, although why this nonplanarity should decrease retention was not clear.) The increasing coefficients in the order $\mathrm{Cl} \leq \mathrm{Br} \leq \mathrm{I}$ indicated that the influences of these halogens was logically through their polarizabilities and the corresponding effect on the dispersion interactions with the stationary phase.[50]

In a study of the gas chromatographic retentions of 206 polychlorinated biphenyls (PCBs) on an SE-54 stationary phase Robbat et al.[51] used six structural parameters to obtain an excellent model ($n = 206$, $r^2 = 0.9989$, $F = 26{,}504$). The parameters used in this study were TWO (number of Cl's at the 2, 2′, 6, and 6′ positions), THREE (Cl's at the 3, 3′, 5, and 5′ positions), FOUR (4 and 4′ Cl's), and ORTHO, META, and PARA to represent intra-ring relationships and interactions.

The parameters employed by Robbat et al. were essentially the same as a set employed earlier by Hale et al.[52] in a study of the GC retentions of 110 (of 135 possible) polychlorinated dibenzofurans (PCDFs) on a DB-5 column. Hale et al. used the positional substitution parameters ONE, TWO, THREE, FOUR and the interactive parameters ORTHO, META, PARA noted above, and two additional descriptors, STERIC and FOURFOUR. STERIC is 1 if both the 2 and 2′ positions are chlorinated, and 0 otherwise.

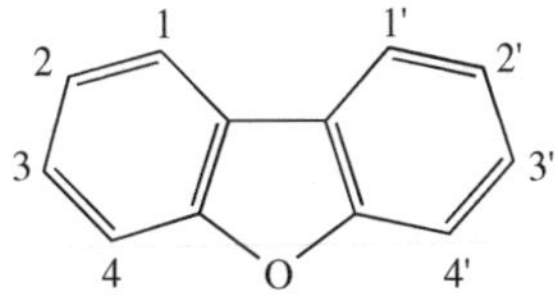

p-Dibenzofuran

FOURFOUR is 1 if the 4 and 4′ positions are occupied, and 0 otherwise. This set of 9 descriptors yielded a model with $n = 110$, $r^2 = 0.99945$, $s = 6.78$, and $F = 21{,}274$. A later examination by Seybold and Hileman[53] of this same set of retentions took a somewhat different approach, focusing first on the simple chlorine count N_{Cl}, and then adding interaction terms.

Taken together, the studies by Hale et al.[52] and Seybold and Hileman[53] shed some interesting light on the factors influencing retention. Seybold and Hileman found that the chlorine count N_{Cl} by itself yielded a model with $r^2 = 0.9838$, $s = 35$, showing the dominant influence of the number of chlorines on the behavior. Addition of two interaction terms, one (VIC) for the number of vicinal (adjacent) Cl pairs and the other (HCL1) for the presence at the 1,1′ positions of an H—Cl pairing, increased r^2 to 0.9956 and reduced s to 18.2. Addition of other interaction terms (for example, for *meta* and *para* relationships, etc.) yielded an 8-term regression of excellent quality ($n = 110$, $r^2 = 0.9995$, $s = 6.59$, $F = 24{,}159$). Also included in the latter study were quantum chemical studies of the chlorine charge variations that helped rationalize some of the forms of the descriptors. Hale et al. found the following regression coefficients for the four chlorine substituent positions:

ONE	TWO	THREE	FOUR
162.3	185.4	180.1	195.7

This indicates that a chlorine substituent at the 4 position is about 20% more effective in prolonging retention than is a chlorine at the 1 position. (The descriptors STERIC and FOURFOUR correct for specific interactions at the 1,1′ and 4,4′ positions.) Since theoretical calculations[53] show that the charges on the chlorines at these positions are essentially the same, one can reasonably conclude that the differing coefficients above reflect positional effects, perhaps related to differences in exposure of substituents at the positions. Note that the presence of a polarizable chlorine atom at the 4 or 4′ position reinforces the charge/polarizability effect of the nearby oxygen lone pair electrons. The latter effect is illustrated in a map of the electrostatic potential near the oxygen atom.[54]

In their study of boiling points, Balaban et al.[38] found that for halomethanes with 2 or more attached halogens an excellent regression was obtained using simple counts of halogens:

$$\mathrm{BP}(^\circ\mathrm{C}) = -38.08\,N_F + 14.60 N_{Cl} + 42.18 N_{Br} + 75.78 N_I + 18.73$$
$$n = 44 \qquad s = 5.26 \qquad r^2 = 0.99 \qquad F = 2204$$

However, when the monohalomethanes were added the correlation degraded considerably and all the monohalomethanes were outliers. In their more extensive examination of 532 C_1—C_4 haloalkanes, these authors found that simple counts alone did not work, whereas addition of the first-order molecular connectivity index ${}^1\chi$ (see below) yielded a regression of very good quality:

$$\mathrm{BP}(^\circ\mathrm{C}) = 138.62\,{}^1\chi - 15.35 N_H - 76.35 N_F - 37.21 N_{Cl} - 14.29 N_{Br}$$
$$+ 14.45 N_I - 53.78 \qquad n = 532 \qquad s = 11.59 \qquad r^2 = 0.97 \qquad F = 2620$$

Correla[55] has also examined the boiling points of the alkyl halides. He notes that some textbooks have tied changes in these boiling points to changes in the molecular weights of the compounds, their surface areas, and/or the halogen polarizabilities. We should note here that increases in both the molecular weights and the surface areas of the halocarbons reflect the increasing size, and increasing polarizabilities, of these atoms in the series $F \leq Cl \leq Br \leq I$. Correla has concluded that induction interactions play a strong role in determining the boiling points of these compounds, with some influence of dispersion interactions for the alkyl bromides and iodides, although the basis for these assertions is not entirely clear. An exception may occur in the alkyl fluorides, where, because of fluorine's low polarizability, polar forces may play a more prominent role.

In a study of general anesthetics in 1977 Di Paolo et al.[56] introduced an empirical "polar hydrogen factor" Q_H based on tabulations of the Pauling electronegativity differences between the bonded halogens and the carbon atom. This empirical correction term often works remarkably well. Using the zeroth-order valence molecular connectivity index ${}^0\chi^v$ (see below for a more detailed discussion) for a set of 28 halogenated gaseous anesthetics Di Paolo et al. found

$$\mathrm{Log}(1/p) = 0.571\,{}^0\chi^v - 0.63$$
$$n = 28 \qquad r = 0.881 \qquad s = 0.496 \qquad F = 90$$

where p is the effective anesthetic pressure in atmospheres of the compounds. Addition of Q_H improved the regression considerably:

$$\mathrm{Log}(1/p) = 0.496\ {}^0\chi^{v} + 10.3\ Q_H - 0.807$$
$$n = 28 \qquad r = 0.966 \qquad s = 0.278 \qquad F = 173$$

Other workers have also used the polar hydrogen factor with success. In a study of eight physical properties—boiling points (BP), melting points (MP), molar refractions (MR), molar volumes (MV), heats of vaporization (HV), surface tensions (ST), critical temperatures (TC), and solubilities log(*S*)—of a set of 25 halogenated methanes, ethanes, and ethylenes, Seybold et al.[57] found that addition of Q_H to counts of the types of carbons and halogens yielded excellent regressions for four of the properties ($r^2 \geq 0.98$, BP, MR, MV, HV), and good regressions ($r^2 \geq 0.90$) for ST and TC. The equation for the boiling points was

$$\mathrm{BP}(^\circ\mathrm{C}) = 29.7N_C - 10.5N_F + 36.0N_{Cl} + 61.1N_{Br} + 20.0Q_H + 178.3$$
$$n = 25 \qquad r^2 = 0.984 \qquad s = 7.2 \qquad p \leq 0.0001$$

Here N_C is the count of carbon atoms ($N_C = 1$ or 2) for this set. In a later study of the tissue (blood, fatty tissue, muscle, and liver) solubilities (represented as tissue/air partition coefficients, log*P*) of this same set of compounds Gargas et al. obtained good ($r^2 = 0.908$–0.978) regressions using carbon and halogen counts and Q_H.[58] Based on results from principal components analysis (PCA), these workers also noted that a good account of these tissue solubilities could be obtained by simply using the measured log*P* values for the compounds in saline and olive oil solutions as descriptors.

Dixon and Seybold[59] have examined the boiling points of a collection of 72 halogenated methanes, ethanes, and ethylenes. Using only the counts of the halogens and an indicator variable (NDB) for the double bond in ethylene they found an equation with $r^2 = 0.957$, $s = 16.2\,^\circ$C, and $F = 237$. Adding Q_H to the regression improved the statistics to $r^2 = 0.985$, $s = 9.7\,^\circ$C, and $F = 586$. At this point the charges on the compounds were calculated using the topological partial equilization of orbital electronegativity (PEOE) method of Gasteiger and Marsili.[25] Taking the charges on the mono-halomethanes as references, the "effective numbers" N_i^{eff} of each of the halogen types were then determined in a manner reflecting the fractional increase or decrease in halogen charge in multiply-substituted halocarbons. Using the adjusted "effective" numbers of halogens with Q_H and NDB improved the regression statistics to $r^2 = 0.992$, $s = 6.94\,^\circ$C, and $F = 1156$. Addition of two parameters related to hydrogen bonding improved the regression statistics further to $r^2 = 0.997$, $s = 4.73\,^\circ$C, and $F = 1946$.

B. Connectivity Indices

In 1975 Randic' introduced the concept of a "branching index" to study the physical properties of saturated hydrocarbons.[60] If one assigns to each carbon in an alkane structure a "valence" δ equal to the number of carbons bonded to it, the branching index is defined as the sum over all C—C bonds i-j of contributions $(\delta_i\delta_j)^{-1/2}$, i.e.

$$\chi = \sum 1/\sqrt{(\delta_i\delta_j)}$$

This idea was later expanded and renamed by Kier and Hall to encompass substructures and to take account of multiple bonds and heteroatoms.[61–64] For substructures of the molecular skeleton, a hierarchy of molecular connectivity indices was defined: e.g. for counts over individual nonhydrogen atoms the zeroth-order molecular connectivity index was defined as

$$^0\chi = \sum 1/\sqrt{\delta_i}$$

and for paths of two bonds each, the second-order connectivity index was defined as

$$^2\chi = \sum 1/\sqrt{(\delta_i\delta_j\delta_k)}$$

Other substructures were included representing longer paths, "clusters", and "path-cluster" combinations.[62,64] Thus the forms $^3\chi_c$, $^4\chi_c$, and $^4\chi_{pc}$ represent the substructure indices for clusters of three bonds, four bonds, and a path-cluster substructure with four bonds, respectively. The preceding superscripts in these terms give the "order" of the index, the following subscripts the type of index. For heteroatoms and multiple bonds, new valence delta values for nonhydrogen atoms were defined as

$$\delta^v = (Z^v - h)/(Z - Z^v - 1)$$

Here Z is the atomic number of the atom, Z^v is the number of valence electrons, and h is the number of attached hydrogen atoms. This form reduces to the ordinary Randic' form for saturated carbon systems. For halogen atoms one finds $\delta^v = 7$ for F, $\delta^v = 7/9$ for Cl. and $\delta^v = 7/27$ for Br. Using these values, valence substructure connectivity indices χ_s^v analogous to the ordinary connectivity indices can be defined.

A very large number of studies have applied molecular connectivity indices (MCIs) to a wide variety of properties. A number of studies before 1985 have been described in the books by Kier and Hall[62,64] and need not be repeated here. Only selected examples from the large number of later studies will be described.

In their extensive study of halocarbon boiling points Balaban et al. derived a number of equations involving different combinations of connectivity indices and halogen counts.[38] An example would be

$$\begin{aligned} BP(^{\circ}C) = {} & 33.21[{}^{1}\chi^{v} - {}^{0}\chi^{v}] - 64.06\,{}^{D}\chi^{0} + 98.46\,{}^{1}\chi - 20.65N_{Br} \\ & - 22.18N_{I} + 6.36[{}^{2}\chi^{v} - {}^{1}\chi^{v}] - 73.65 \\ & n = 532 \quad r^{2} = 0.97 \quad s = 10.9 \quad F = 2953 \end{aligned}$$

Seybold and coworkers employed connectivity indices in their examination of eight physical properties of C_1 and C_2 halocarbons.[57] For the properties BP, MR, MV, HV, and TC these indices yielded regressions with $r^2 > 0.95$. For example, for the heats of vaporization,

$$\begin{aligned} HV(kJ/mol) = {} & 21.6\,{}^{1}\chi^{v} + 6.07\,{}^{1}\chi - 2.90\,{}^{0}\chi^{v} + 3.50 \\ & n = 17 \quad r^{2} = 0.990 \quad s = 0.93 \quad p, 0.0001 \end{aligned}$$

In a study of the tissue solubilities of this set of compounds connectivity indices were only moderately successful ($r^2 = 0.85$–0.88 for blood, liver, and muscle, $r^2 = 0.978$ for fatty tissue) in representing the tissue/air partition constants (as Log*P*s), but combinations of the connectivity indices with halogen counts and Q_H gave $r^2 > 0.939$ for all the tissues.[58]

Robbat et al.[65] employed MCIs to analyze the reported gas chromatographic retentions of polychlorinated biphenyls on an SE-54 stationary phase. For a reported set of 206 PCB congeners they achieved an excellent regression using a set of five connectivity indices (${}^{4}\chi_{pc}, {}^{5}\chi, {}^{2}\chi^{v}, {}^{5}\chi^{v}_{c}, {}^{5}\chi^{v}$: $n = 206, r^2 = 0.9987, s = 0.0064, F = 31057$).

Katritzky and Gordeeva[66] analyzed the performances of several types of molecular descriptors in estimating a variety of different physical (melting point, boiling point, refractive index, molar volume, and density) and biological (anaesthetic activity, narcotic activity, and sweetness intensity) properties. They tested 84 descriptors and determined which participated most often in the "best" regression models. Among the descriptors tested they found what they called a "clear champion," the classic first-order Randic' connectivity index ${}^{1}\chi$. Second and third places went to ${}^{3}\chi$ and ${}^{1}\chi^{v}$.

Sabljic' and coworkers have employed MCIs in a number of applications, especially environmentally related issues such as bioconcentration,[67,68] soil sorption,[69–71] and biodegradability.[72] They have also used MCIs to model chromatographic retention indices for chlorinated compounds,[73,74] toxicities of chlorinated compounds,[75,76] and reaction rates.[77,78]

C. Optimal Descriptors

Randic' has proposed a scheme ("optimal moleculsr descriptors", OMD) in which one employs a limited set of descriptors and attempts to optimize these.[79,80] Specifically, one takes the adjacency matrix (the matrix of the hydrogen-suppressed graph in which bonds between atoms are assigned values of 1 and the diagonal terms represent the appropriate values for the atoms in question) and assigns variable terms, e.g. x, y, ..., to the different types of atoms. This scheme is analogous to the treatment of heteroatoms in the Hückel molecular orbital (HMO) method.[81] For example, the OMD approach replaces the diagonal terms of the normal adjacency matrix for ethyl alcohol with a "generalized adjacency matrix":

$$\begin{pmatrix} 0 & 1 & 0 \\ 1 & 0 & 1 \\ 0 & 1 & 0 \end{pmatrix} \quad \text{replaced by} \quad \begin{pmatrix} x & 1 & 0 \\ 1 & x & 1 \\ 0 & 1 & y \end{pmatrix}$$

where x and y are variables for the carbon and oxygen atoms, respectively. In a study of the boiling points of alcohols, Randic' found that values of $x = 1.50$ and $y = -0.85$ yielded the smallest standard error.

Randic' has applied this approach[79] to the biological activities of a set of 18 clonidine-like imidazolines taken from an earlier study by Timmermans and van Zweiten:[82]

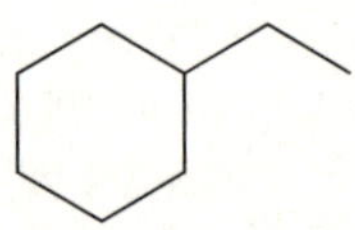

Clonidine skeleton

These earlier workers employed a variety of descriptors, both experimental and computed, to achieve a regression with $r = 0.95$, $s = 0.35$ for a more diverse set of 27 compounds. (As early as 1965 Charton[83] found a significant relationship between the pK_a's of imidazoles and Hammett constants, and more recently Mitchell et al.[84] and Kim and Martin[85] have obtained excellent models for the pK_a's of clonidine analogs using HOMO energies and nitrogen charges, and CoMFA analysis and Hammett constants, respectively.) The 18 compounds studied by Randic' were those with chlorine as the only heteroatom present. Taking $x = -0.40$ for chlorine and calculating the "weighted" connectivity indices ${}^1\chi', {}^2\chi'$, and ${}^3\chi'$, Randic' achieved an equation with $r = 0.977$ and $s = 0.22$ for the antihypertensive activities of these compounds. Recently Randic' and Basak have applied this same technique in several additional areas.[86,87]

Abraham et al.[29] have used their atomic charges, calculated empirically, to estimate the dipole moments of fluoroalkanes, chloroalkanes, bromoalkanes, iodoalkanes and mixed halocarbons with considerable success. These workers also studied a number of conjugated and aromatic compounds with F, Cl, Br and I substituents.[30]

D. Diverse Descriptors

By far the largest number of QSAR and QSPR studies of halogenated compounds have employed a variety of different descriptors from several categories. Katritzky and coworkers[88,89] have identified the most common molecular structure descriptor categories as (1) constitutional (depending on the composition of the molecule); (2) topological; (3) electrostatic (partial charges and electronegativities); (4) geometric (three-dimensional structure); and (5) quantum chemical (net charges, MO energies, etc.).

A good example of this approach is the study by Katritzky et al.[90] of the melting points of a set of 443 mono- and disubstituted benzenes, which included a large number of halogenated benzenes. These workers developed a 9-parameter equation with $r^2 = 0.837$ for the melting points, with the most significant descriptor being the hydrogen donor charged surface area (HDSA) determined using the charges calculated by the method of Zefirov.[91] The second most significant descriptor was the average valency of the hydrogen atoms on the compound.

Huibers and Katritzky[92] studied the aqueous solubilities of a set of 109 hydrocarbons and 132 halogenated hydrocarbons using a variety of descriptors. In this study they found the molar volume to be the key descriptor, as demonstrated by the single-descriptor relation

$$-\log S_w = -1.27 + 0.0380\,\mathrm{MV}$$

$$n = 241 \qquad r^2 = 0.904 \qquad s = 0.590 \qquad F = 2253$$

Including two additional descriptors brought this to $r^2 = 0.959$, $s = 0.386$.

The research efforts of the Jurs group have traditionally included a variety of descriptors from different categories. For the vapor pressures (VPs, in Pascals) of a collection of 352 hydrocarbons and halohydrocarbons they found a 7-descriptor model for log(VP) with $r^2 = 0.983$ and $s = 0.186$.[93] The descriptors in the model included molecular connectivity indices, atom and bond counts, and descriptors related to the charged partial surface areas. A study of the aqueous solubilities of a diverse set of organic compounds including many chlorinated examples used a similarly diverse set of descriptors.[94] Other studies have concerned solubilities in supercritical carbon dioxide,[95] infinite dilution activity coefficients,[96] aqueous solubilities,[97] critical temperatures and pressures,[98] and radical reaction rate constants.[99,100]

E. Gravitation Index

In 1996 Katritzky et al.[101] introduced a descriptor which they termed the "gravitation index" and defined as

$$G_1 = \sum m_i m_j / r_{ij}^2$$

where the m's are the atomic masses, r_{ij} is the distance between atoms i and j, and the summation is carried out over all bonded pairs of atoms. For the boiling points of a set of 298 diverse organic compounds they used $(G_1)^{1/3}$ with a hydrogen-bonding term (the area-weighted surface charge of the hydrogen-bonding donor atoms) to obtain an expression with $r^2 = 0.9544$ and $s = 16.2\,\mathrm{K}$. They have used the gravitaion index extensively in their subsequent work. In a study of the critical temperatures of 76 compounds they found the single-parameter model[102]

$$T_c = -100 + 79.7(G_1)^{1/3}$$

$$n = 76 \qquad r^2 = 0.953 \qquad s = 18.9\,\mathrm{K} \qquad F = 1506$$

In a study of the vapor pressures, aqueous solubilities, and water/air partition coefficients of a diverse set of 411 compounds Katritzky et al.[103] found that for the vapor pressures G_1 was the most important descriptor in a 5-parameter equation with $r^2 = 0.949$. G_1 was also used in a study of the refractive indices of a diverse set of 125 organic compounds[104] and in a study of the liquid viscosities of a diverse set of 361 organic compounds, including a number of halogenated compounds.[105]

F. Quantum Chemical Descriptors

Pure quantum chemical QSAR and QSPR treatments of halogenated compounds are relatively rare, although a number of studies have used quantum chemical calculations in combination with other descriptors, as in the uses of charged partial surface areas (CPSAs) cited below. The use of quantum chemical descriptors in QSAR and QSPR has been reviewed by Karelson et al.[106] A very early example of the use of quantum chemical calculations in this area is the study of anesthetic action by Agin et al.[107] These workers employed ionization potentials I_p estimated using Hückel molecular orbital theory with available polarizability (α) values (taken as molecular values or sums of atomic values) to estimate the minimum blocking concentration (MBC) for a collection of 39 compounds.

Considering the simplicity of the calculations, a surprisingly good linear correlation was obtained (no statistics cited) between the measured log(MBC) values and the product term αI_p.[107] A review of "MO-QSARs" prior to 1990 has been given by Lewis.[108]

A more recent study of the chromatographic retentions of PCBs by Ong and Hites demonstrates the potential of the pure quantum chemical approach.[48] These workers employed Dewar's semiempirical AM1 molecular orbital method[109] to calculate the molecular polarizabilities, ionization potentials, and dipole moments of polycyclic aromatic hydrocarbons (PAHs), polychlorinated biphenyls (PCBs), polychlorinated dibenzo-*p*-dioxins (PCDDs), and polychlorinated dibenzofurans (PCDFs). The PCDD set examined was limited to congeners with four or more chlorines. Under the assumption that the GC retentions of these compounds are mediated through van der Waals forces, namely London dispersion forces (dependent upon I_p and α), induction forces (dependent upon α), and orientation forces (proportional to the square of the dipole moment, μ^2), they derived a model for the retention indices of the following form:

$$\mathrm{RI} = b_0 + b_1\alpha + b_2 I_p + b_3\mu^2 \quad (1)$$

Before attempting to model the four classes of compounds Ong and Hites ascertained, as well as possible, that the AM1 method gave calculated values for α, I_p, and μ that were in satisfactory agreement with the limited number of experimental values available for these compounds. Using the equation above they obtained impressive results for all four classes of compound, with r^2 values ranging from 0.927 for the PCDFs to 0.987 for the PAHs. Their regression results are summarized in Table 2. These results further allow some reasonable inferences regarding the interactions responsible for the retentions. For all four models the molecular polarizability accounted for more than 90% of the variance, suggesting that the main operational forces are dispersion forces. This result is consistent with earlier studies by Meyer and coworkers[110] on the forces prevailing in liquids, which showed

Table 2. Regression coefficients and statistics from the study of Ong and Hites.[48] The equation is $\mathrm{RI} = b_0 + b_1\alpha + b_2 I_p + b_3\mu^2$. [$\alpha$ in Å^3, I_p in eV, μ^2 in Debyes^2]

Type	b_0	b_1	b_2	b_3	*n*	r^2	*s*	*F*
PAHs	−1250	91.91	141.22	−11.31	63	0.987	59.1	1438
PCBs	473	149.09	−163.36	13.35	209	0.964	52.4	1824
PCCDs	−3505	149.17	275.30	15.25	37	0.981	32.2	575
PCDFs	−3085	141.58	247.17	2.56	69	0.927	55.7	274

that even for relatively polar molecules such as acetonitrile, chloromethane, and acetone, dispersion forces provide most of the cohesive energy. Similar conclusions were reached by Seybold and Bertrand in their study of the chromatographic retentions of polyhalogenated biphenyls.[50]

The study by Ong and Hites shows the potential power of the pure quantum chemical approach, and should encourage further studies of this nature. An advantage of this method is that from a very basic approach one can infer important conclusions about the fundamental causes of the phenomenon investigated.

Katritzky, Karelson and their coworkers[111] used a combination of quantum chemical descriptors along with conventional descriptors to analyze the gas chromatographic retention times of a diverse group of 152 compounds. They employed the AM1 semiempirical method[109] for the quantum chemical calculations. For the retention times they found a two-parameter equation with $r^2 = 0.913$ and a six-parameter equation with $r^2 = 0.959$. The most important descriptors were the calculated polarizability and the minimum valency on an H atom, describing the dispersion and hydrogen-bonding interactions, respectively.

Politzer, Murray, and their coworkers have used aspects of the electrostatic potentials of compounds to study a number of properties of organic compounds, including reactivities and pK_as.[112–115] For group V–VII hydrides and their corresponding anions they found high-quality ($r^2 = 0.962$–0.991) regression equations for enthalpies of protonation and pK_as using variables such as the minimum of the electrostatic potential (V_{min}) and the minimum of the local ionization potential ($I_{S,min}$).[115] Gross and Seybold[34,35] studied the pK_as of substituted anilines and phenols using quantum chemical charge and energy descriptors, and also found strong correlations. For the substituted phenols[35] they found the following relations:

$$pK_a = -38.4\, Q_n(O^-) - 19.0$$

$$n = 19 \qquad r^2 = 0.911 \qquad s = 0.25 \qquad F = 173$$

$$pK_a = 49.5\, E_{HOMO} + 13.8$$

$$n = 19 \qquad r^2 = 0.951 \qquad s = 0.19 \qquad F = 328$$

Here $Q_n(O^-)$ is the natural charge on the oxide oxygen atom and E_{HOMO} is the energy of the highest occupied MO.

The main avenue of approach thus far has been to combine quantum chemical parameters with other descriptors. For example, Shusterman and coworkers have combined octanol/water partition coefficients, as log*P*s, with quantum chemical descriptors such as HOMO and LUMO energies to examine biological activities.[116,117] For the mutagenic activity (Ames test) of nitro aromatic compounds against the *S. Typhimurium* strain TA100 the following regression equation was found:

$$\log(\mathrm{TA100}) = 1.36 \log \mathrm{P} - 1.98\, E_{\mathrm{LUMO}} - 7.01$$
$$n = 47 \qquad r = 0.911 \qquad s = 0.737$$

G. Charged Partial Surface Areas

Many molecular properties depend on polar interactions either among the molecules themselves or between the molecules and other entities such as stationary phases in chromatography or receptors in drug-receptor interactions. Polar interactions can be especially important in some cases for halogenated compounds, particularly those with fluorine substituents. One effective way to treat such interactions is through charged partial surface area (CPSA) descriptors, a method developed by Grigoras[118] and Stanton and Jurs.[119] In the approach adopted by Stanton and Jurs, the molecular surface is characterized by the overlap of van der Waals spheres for the atoms. From this, the solvent-accessible surface[132] is determined (using a spherical water molecule) and portions assigned to the different atoms. Atomic charges were determined using a method similar to that of Abraham and Smith.[30] These values were combined in several ways to create descriptors encoding the information in different ways, defining, e.g. the partial positive and negative surface areas, PPSA and PNSA, which simply are sums of the positive and negative areas, charge-weighted surface areas, and fractional and relative areas. These authors demonstrated the utility of the CPSA indices with applications to the chromatographic retentions of a set of 107 pyrazines, the boiling points of 209 furan and tetrahydrofuran related compounds, and the surface tensions of a diverse set of 31 organic compounds.[119]

The Jurs group has further applied these descriptors in a large number of applications.[120–124]

H. Surface Areas and Volumes

Molecular volumes and surface areas have been employed in many studies.[125,132] Bhattacharjee, for example, has used the geometric volume as a descriptor in studies of a large number of properties of halomethane[126] and haloethane[127] refrigerants. By "geometric volume" this author means "the volume of the geometric shape of the unit molecule assuming atoms as point masses."[126] In this way Bhattacharjee obtained linear and parabolic regressions for melting points, critical properties (T_c, P_c, V_c), viscosity, refractive index, thermal conductivity, saturated vapor density, dielectric constant, surface tension and critical density. Parabolic regressions with r^2

values from 0.938 to 0.999 were obtained for the halomethanes, and from 0.826 (melting points) to 0.989 for the haloethanes. Molar volumes often show strong correlations with a number of these properties.[128]

Karelson and Perkson[129] have analyzed the densities of a set of 61 halocarbons using a single descriptor, the intrinsic density ρ^R, determined as the ratio of the calculated van der Waals volume (with overlapping spheres) to the mass:

$$\rho = 0.8706\,\rho^R + 0.0769$$
$$n = 61 \qquad r^2 = 0.946 \qquad s = 0.085 \qquad F = 1042$$

For a larger diverse set of 303 compounds containing C, H, O, N, S, F, Cl, Br, and I they added a term $E_{el.stat}$ for the molecular electrostatic interaction per atom to achieve a regression with $r^2 = 0.9749$, $s = 0.046\,g/cm^3$, and $F = 5937$.

Several studies already mentioned have employed the molecular surface areas of compound in correlations with different properties.[45,55] Doucette and Andren[130] studied the octanol/water partition coefficients of halogenated biphenyls, dibenzodioxins and dioxins using the total surface area (TSA) of each compound as a parameter. De Bruijn and Hermens[131] looked at the octanol/water partition coefficients of chlorinated benzenes and biphenyls using both the solvent accessible molecular surface area and the total molecular volume as parameters, finding good results. Molecular surface areas and volumes—the two are often strongly correlated[132]—are especially useful in examining properties such as solubility,[133,134] partition coefficients,[135] and chromatographic retentions,[130,136,137] since the cavity term plays a significant role in these properties. Dunnivant et al.[138] used TSA with melting points (MP) and a "shadow area" (SA) descriptor to achieve an excellent expression for the solubilities of a set of PCBs:

$$\log S + -0.0352\,\text{TSA} - 0.0103\,\text{MP} - 0.0389\,\text{SA} + 8.62$$
$$n = 22 \qquad r = 0.992 \qquad s = 0.165 \qquad F = 329$$

Yalkowsky and Valvani[139] had earlier demonstrated that expressions of the form $\log S = a(\text{TSA}) + b(\text{MP})$ were useful for solubilities. Amidon and Anik[134] have noted that a number of popular topological indices, including the Wiener number,[140] Hosoya's Z value,[141] and Randic's connectivity index,[60] tend to correlate strongly with molecular surface area.

I. Electrotopological States

We saw in the Introduction that the activity of a halogen atom in intermolecular interactions is generally a composite result of the atom's availability to interact with other molecules (exposure) and its electronegativity and polarizability as mediated through its electronic charge. Just over a decade ago Kier and Hall introduced in a series of papers[142–145] a new atomic descriptor, the electrotopological state index S, designed to take into account both of these influences. Work with this descriptor has recently been summarized in a book.[146]

Construction of the electrotopological index begins by assigning to each type of atom or hydride group (atom plus associated hydrogens) i an "intrinsic state" index I_i, defined, for second-row atoms, as

$$I_i = (\delta^v + 1)/\delta$$

where δ^v and δ are the valence and ordinary delta values defined earlier in the context of molecular connectivity indices. The value δ^v represents the number of sigma, pi, and lone pair valence electrons on the atom, whereas δ encompasses only the number of sigma electrons involved in bonds to nonhydrogen neighbors. For atoms in higher rows of the periodic table a more general expression is

$$I_i = \{(2/N)^2\delta^v + 1\}/\delta$$

where N is the principal quantum number of the valence level. Account of the influence of other atoms in the molecular structure on the atom's state is accomplished by including perturbation terms of the form

$$\Delta_{ij} = (I_i - I_j)/R_{ij}^{m}$$

where I_i and I_j are the intrinsic states of the two atoms, m is an integer exponent, and R_{ij} is a measure of the topological "distance" between the two atoms, taken as the number of atoms in the shortest path linking the two atoms (and equal to 1 + the number of bonds in the path). The topological state index for atom i is then defined as

$$S_i = I_i + \sum_j \Delta_{ij}$$

In this way the index accounts for both the electronic nature of the atom and its accessibility for interactions with other atoms.

The intrinsic state I_i values for the halogens and for some representative forms of carbon are as follows:

$—CH_3$	>C=	$—CH_2—$	F	Cl	Br	I
2.000	2.000	1.500	8.000	4.111	2.750	2.120

Kier and Hall[147] noted quite some time ago that the values calculated in the above manner correlate with Mulliken–Jaffe electronegativity values.[148] Joshi[146,149] has compared the topological state values with those of a number of other topological indices and found a considerable degree of independence in most cases. There does, however, occur a strong correlation with "free valence indices" obtained from the Hückel molecular orbital method.

The central question, of course, is whether these indices are useful in examining molecular properties in QSAR and QSPR. It appears that they are. In a study of the rates of hydrogen abstraction by 26 chlorofluorocarbons Kier and Hall found the equation[150]

$$\log k = 0.377\, S_1 - 1.346\, {}^3\chi_c + 0.671\, {}^1\kappa_\alpha + 5.415$$

$$n = 26 \qquad r^2 = 0.89 \qquad s = 0.27 \qquad F = 60$$

where S_1 is the E-state index for the carbon atom with the greatest number of bound halogens, ${}^3\chi_c$ is the third-order cluster molecular connectivity index, and ${}^1\kappa_\alpha$ is Kier's shape index. In a study of the toxicities of 141 substituted benzenes, including a number of halogenated compounds, toward fathead minnows (*Pimephales promelas*), Kier and Hall[146] used E-state indices as inputs for a neural network examination. They achieved $r^2 = 0.86$ for the training set and $r^2 = 0.89$ for the test set in this study. In another neural network study of the boiling points of a mixed set of 298 compounds, some with chlorines or bromines, they found $r^2 = 0.998$ for the total set. These authors also describe how the E-state indices can be employed to reveal patterns of information, taking the polychlorinated biphenyls as an illustration.[146] They create a space defined by two axes, one the E-state total for the chlorines in a PCB congener, the other the E-state total for the aromatic CH groups. When the congeners are plotted in this space, they fall naturally into groups according to their numbers of chlorines; if these groups are more closely examined they reveal further patterns separating the compounds into subgroups according to the positional variations present. Kier and Hall also note the potential of E-states for aiding in other applications such as database searching and similarity analysis.[146]

IV. CONCLUSIONS

It is apparent that there are a considerable number of methods currently available for treating halogens in QSAR and QSPR studies, despite the potential complexity of the electronic behaviors of these atoms. The important roles, both favorable and unfavorable, of halogen compounds in commercial applications, as drugs, and as environmental contaminants suggest that this area will continue to develop and evolve new methods.

ACKNOWLEDGMENTS

The author thanks Drs. Sandy Balaban, Lowell Hall, Peter Jurs, Mati Karelson, and Peter Politzer for making reprints of their work on this topic available to him. He also thanks Dr. Lemont B. Kier and Mr. Kevin Gross for helpful discussions of this topic, and Kevin Gross for preparing Figure 1.

REFERENCES

1. Cramer, R. D. III; Patterson, D. E; Bunce, J. D. *J. Am. Chem. Soc.* **1988**, *110*, 5959–5967.
2. Tosato, M. L.; Marchini, S.; Passerini, L.; Pino, A.; Eriksson, L; Jonsson, J.; Hellberg, S.; Lindgren, F.; Skagerberg, B; Sjöström, M.; Wold, S.; Berglund, R. *Environ. Toxicol. Chem.* **1990**, *9*, 265–277.
3. Lindgren, F.; Eriksson, L.; Hellberg, S.; Jonsson, J.; Sjöström, M.; Wold, S. *Quant. Struct.-Act. Relat.* **1991**, *10*, 36–42.
4. Zhang, D. *Comput. Chem.* **2001**, *25*, 197–204.
5. McMurray, *Organic Chemistry*, 5th Ed., Brooks/Cole, **1999**.
6. Politzer, P. *J. Am. Chem. Soc.* **1969**, *91*, 6235–6237.
7. Politzer, P.; Timberlake, J. W. *J. Org. Chem.* **1972**, *37*, 3557–3559.
8. Applequist, J.; Carl, J. R.; Fung, K.-K. *J. Am. Chem. Soc.* **1972**, *94*, 2952–2960.
9. Applequist, J. *Accts. Chem. Res.* **1977**, *10*, 79–85.
10. Bode, K.; Applequist, J. *J. Phys. Chem.* **1996**, *100*, 17820–17824.
11. Le Fevre, R. J. W. In *Advances in Physical Organic Chemistry*, Vol. 3, Gold, V. Ed., Academic Press: London, **1965**, p. 1.
12. Politzer, P. *J. Chem. Phys.* **1987**, *86*, 1072–1073.
13. Huheey, J. E. *J. Org. Chem.* **1971**, *36*, 204–205.
14. Watts, J. C.; Huheey, J. E. *Chem. Phys. Lett.* **1972**, *14*, 89–90.
15. Filgueiras, C. A. L.; Huheey, J. E. *J. Org. Chem.* **1976**, *41*, 49–53.
16. Huheey, J. E. *Inorganic Chemistry*, 3rd Ed., Harper and Row:, New York, **1983**, Chap. 3.
17. Williams, D. E. In *Revs. Comput. Chem.*, Lipkowitz, K. B.; Boyd, D. B. Eds., Vol. II, **1991**, pp. 219–271.
18. Bachrach, S. M. In *Revs. Comput. Chem.*, Lipkowitz, K. B.; Boyd, D. B. Eds., Vol. V, **1994**, pp. 171–227.
19. Wiberg, K. B.; Hadad, C. M.; Breneman, C. M.; Laidig, K. E.; Murcko, M. A.; LePage, T. J. *Science* **1991**, *252*, 1266–1272.
20. Rappé, A. K.; Goddard, W. A. III. *J. Phys. Chem.* **1991**, *95*, 3358–3363.
21. Reed, A. E.; Weinstock, R. B.; Weinhold, F. *J. Chem. Phys.* **1985**, *83*, 735–746.
22. Weinhold, F. In *Encyclopedia of Computational Chemistry*, Wiley: New York, Vol. 3, pp. 1792–1811.
23. Bader, R. F. W. *Accts. Chem. Res.* **1985**, *18*, 9–15.
24. Bader, R. F. W. *Atoms in Molecules—A Quantum Theory*, Oxford Univ. Press, Oxford, U.K., **1990**.
25. Gasteiger, J; Marsili, M. *Tetrahedron* **1980**, *36*, 3219–3228.
26. Mortier, W. J.; Van Genechten, K.; Gasteiger, J. *J. Am. Chem. Soc.* **1985**, *107*, 829–835.

27. No, K. T.; Grant, J. A.; Scheraga, H. A. *J. Phys. Chem.* **1990**, *94*, 4732–4739.
28. Mullay, J. *J. Am. Chem. Soc.* **1986**, *108*, 1770–1775.
29. Abraham, R. J.; Griffiths, L.; Loftus, P. *J. Comput. Chem.* **1982**, *3*, 407–416.
30. Abraham, R. J.; Smith, P. E. *J. Comput. Chem.* **1987**, *9*, 288–297.
31. Hammarstrom, L.-G.; Liljefors, T.; Gasteiger, J. *J. Comput. Chem.* **1988**, *9*, 424–440.
32. Park, J. M.; No, K. T.; Jhon, M. S.; Scheraga, H. A. *J. Comput. Chem.* **1993**, *14*, 1482–1490.
33. Dixon, S. L.; Jurs, P. C. *J. Comput. Chem.* **1992**, *13*, 492–504.
34. Gross, K. C.; Seybold, P. G. *Int. J. Quantum Chem.* **2000**, *80*, 1107–1115.
35. Gross, K. C.; Seybold, P. G. *Int. J. Quantum Chem.* **2001**, *85*, 569–579.
36. Gross, K. C.; Seybold, P. G.; Hadad, C. *Int. J. Quantum Chem.* **2002** (in press).
37. Palmer, M. H. *J. Mol. Struct. (Theochem)* **2000**, *500*, 225–243.
38. Balaban, A. T.; Joshi, N.; Kier, L. B.; Hall, L. H. *J. Chem. Info. Comput. Sci.* **1992**, *32*, 233–237.
39. Gupta, S. P. *Chem. Revs.* **1989**, *89*, 1765–1800.
40. Lawson, D. D.; Maocanin, J.; Scherer, K. V.; Terranova, T. F.; Ingham, J. D. *J. Fluorine Chem.* **1978**, *12*, 221–236.
41. Kelly, C. M.; Mathias, P. M.; Schweighardt, F. K. *Ind. Engr. Chem. Res.* **1988**, *27*, 1732–1736.
42. Woolf, A. A. *J. Fluorine Chem.* **1993**, *63*, 19–24.
43. Van Der Puy, M. *J. Fluorine Chem.* **1993**, *63*, 165–172.
44. Yamanouchi, K.; Heldebrant, C. *CHEMTECH* **1992**, 354–359.
45. Höfler, F.; Melzer, H.; Möckel, H. J.; Robertson, L. W.; Anklam, E. *J. Agric. Food. Chem.* **1988**, *36*, 961–965.
46. Recker, R. F.; De Vries, G.; Bijloo, G. J. *J. Liq. Chromatogr.* **1989**, *12*, 695–728.
47. Hasan, M. N.; Jurs, P. C. *Anal. Chem.* **1990**, *62*, 2318–2323.
48. Ong, V. S.; Hites, R. A. *Anal. Chem.* **1991**, *63*, 2829–2834.
49. Kaliszan, R. *Anal. Chem.* **1992**, *64*, 619A–631A.
50. Seybold, P. G.; Bertrand, *J. Anal. Chem.* **1993**, *65*, 1631–1634.
51. Robbat, A., Jr.; Xyrafas, G.; Marshall, D. *Anal. Chem.* **1988**, *60*, 982–985.
52. Hale, M. D.; Hileman, F. D.; Mazer, T.; Shell, T. L.; Noble, R. W.; Brooks, J. J. *Anal. Chem.* **1985**, *57*, 640–648.
53. Seybold, P. G.; Hileman, F. D. (unpublished results).
54. Murray, J. S.; Zilles, B. A.; Jayasuriya, K.; Politzer, P. *J. Am. Chem. Soc.* **1986**, *108*, 915–918.
55. Correla, J. *J. Chem. Educ.* **1988**, *65*, 62–64.
56. Di Paolo, T.; Kier, L. B.; Hall, L. H. *Molec. Pharm.* **1977**, *13*, 31–37.
57. Seybold, P. G.; May, M. A.; Gargas, M. L. *Acta Pharm. Yugosl.* **1986**, *36*, 253–265.
58. Gargas, M. L.; Seybold, P. G.; Andersen, M. E. *Toxicol. Lett.* **1988**, *43*, 235–256.
59. Dixon, S. L.; Seybolf, P. G. (unpublished results).
60. Randic', M. *J. Am. Chem. Soc.* **1975**, *97*, 6609–6615.
61. Kier, L. B.; Hall, L. H. *J. Pharm. Sci.* **1976**, *65*, 1806–1809.
62. Kier, L. B.; Hall, L. H. *Molecular Connectivity in Chemistry and Drug Research*, Academic Press: New York, **1976**.
63. Kier, L. B.; Hall, L. H. *J. Pharm. Sci.* **1983**, *72*, 1170–1173.
64. Kier, L. B.; Hall, L. H. *Molecular Connectivity in Structure-Activity Analysis*, Wiley: New York, **1986**.
65. Robbat, A., Jr.; Xyrafas, G.; Marshall, D. *Anal. Chem.* **1988**, *60*, 982–985.
66. Katritzky, A. R.; Gordeeva, E. V. *J. Chem. Info. Comput. Sci.* **1993**, *33*, 835–857.
67. Sabljic, A.; Protic, M. *Chem.-Biol. Interact.* **1982**, *42*, 301–310.
68. Sabljic, A.; Piver, W. T. *Environ. Toxicol. Chem.* **1992**, *11*, 961–972.

69. Sabljic, A.; Protic, M. *Bull. Environ. Contam. Toxicol.* **1982**, *28*, 162–165.
70. Sabljic, A. *J. Agr. Food Chem.* **1984**, *32*, 243–246.
71. Sabljic, A. *Environ. Sci. Technol.* **1987**, *21*, 358–366.
72. Sabljic, A.; Güsten, H. *Chemosphere* **1989**, *19*, 1503–1511.
73. Sabljic, A. *J. Chromatogr.* **1984**, *314*, 1–12.
74. Sekusak, S.; Sabljic, A. *J. Chromatogr.* **1993**, *628*, 69–79.
75. Sabljic, A. *Bull. Environ. Contam. Toxicol.* **1983**, *30*, 80–83.
76. Protic, M.; Sabljic, A. *Aquatic Toxicol.* **1989**, *14*, 47–64.
77. Sabljic, A.; Güsten, H. *Atmospheric Environ.* **1990**, *24A*, 73–78.
78. Sekusak, S.; Liedl, K. R.; Sabljic, A. *J. Phys. Chem.* **1998**, *102*, 1583–1594.
79. Randic', M. *Chemometr. & Intel. Lab. Syst.* **1991**, *10*, 213–227.
80. Randic', M. *J. Comput. Chem.* **1991**, *12*, 970–980.
81. Streitwieser, A. Jr. *Molecular Orbital Theory for Organic Chemists*, Wiley: New York, **1961**, Chap. 5.
82. Timmermans, B. M. W. M.; van Zwieten, P. A. *J. Med Chem.* **1977**, *20*, 1636–1644.
83. Charton, M. *J. Org. Chem.* **1965**, *30*, 3346–3350.
84. Mitchell, T. J.; Tute, M. S.; Webb, G. A. A. *Eur. J. Med. Chem.* **1990**, *25*, 117–120.
85. Kim, K. H.; Martin, Y. *J. Med. Chem.* **1991**, *34*, 2056–2060.
86. Randic', M.; Basak, S. C. *J. Chem. Info. Comput. Sci.* **1999**, *39*, 261–266.
87. Randic', M.; Basak, S. C. *SAR and QSAR in Environ. Res.* **2000**, *11*, 1–23.
88. Murugan, M.; Grendze, M. P.; Toomey, J. E.; Katritzky, A. R.; Karelson, M.; Lobanov, V. S.; Rachwal, P. *CHEMTECH* **1994**, (June) 17–23.
89. Katritzky, A. R.; Lobanov, V. S.; Karelson, M. *Chem. Soc. Revs.* **1995**, 279–287.
90. Katritzky, A. R.; Maran, U.; Karelson, M.; Lobanov, V. S. *J. Chem. Info. Comput. Sci.* **1997**, *37*, 913–919.
91. Zefirov, N. S.; Kirpichenko, M. A.; Izmailov, F. F.; Trofimov, M. I. *Dokl. Akad, Nauk SSSR* **1987**, *296*, 883–887.
92. Huibers, P. D. T.; Katritzky, A. R. *J. Chem. Info. Comput. Sci.* **1998**, *38*, 283–292.
93. Goll, E. S.; Jurs, P. C. *J. Chem. Info. Comput. Sci.* **1999**, *39*, 1081–1089.
94. Sutter, J. M.; Jurs, P. C. *J. Chem. Info. Comput. Sci.* **1999**, *86*, 100–107.
95. Englehardt, H. L.; Jurs, P. C. *J. Chem. Info. Comput. Sci.* **1997**, *37*, 478–484.
96. Mitchell, B. E.; Jurs, P. C. *J. Chem. Info. Comput. Sci.* **1998**, *38*, 200–209.
97. Mitchell, B. E.; Jurs, P. C. *J. Chem. Info. Comput. Sci.* **1998**, *38*, 489–496.
98. Turner, B. E.; Costello, C. L.; Jurs, P. C. *J. Chem. Info. Comput. Sci.* **1998**, *38*, 639–645.
99. Bakken, G. A.; Jurs, P. C. *J. Chem. Info. Comput. Sci.* **1999**, *39*, 508–514.
100. Bakken, G. A.; Jurs, P. C. *J. Chem. Info. Comput. Sci.* **1999**, *39*, 1064–1075.
101. Katritzky, A. R.; Mu, L.; Lobanov, V. S.; Karelson, M. *J. Phys. Chem.* **1996**, *100*, 10400–10407.
102. Katritzky, A. R.; Mu, L.; Karelson, M. *J. Chem. Info. Comput. Sci.* **1998**, *38*, 293–299.
103. Katritzky, A. R.; Wang, Y.; Sild, S.; Tamm, T.; Karelson, M. *J. Chem. Info. Comput. Sci.* **1998**, *38*, 720–725.
104. Katritzky, A. R.; Sild, S.; Karelson, M. *J. Chem. Info. Comput. Sci.* **1998**, *38*, 840–844.
105. Katritzky, A. R.; Chen, K.; Wang, Y.; Karelson, M.; Lucic, B.; Trinajstic, N.; Suzuki, T.; Schüürmann, G. *J. Phys. Org. Chem.* **2000**, *13*, 80–86.
106. Karelson, M.; Lobanov, V. S.; Katritzky, A. R. *Chem. Revs.* **1996**, *96*, 1027–1043.
107. Agin, D.; Hersch, L.; Holtzman, D. *Proc. Nat. Acad. Sci. (USA)* **1965**, *53*, 952–958.
108. Lewis, D. F. W. *Progr. in Drug Metab.* **1990**, *12*, 205–255.
109. Dewar, M. J. S.; Zoebisch, E. G.; Healy, E. F.; Stewart, J. J. P. *J. Am. Chem. Soc.* **1985**, *107*, 3902–3909.
110. Meyer, E. F.; Renner, T. A.; Stee, K. S. *J. Phys. Chem.* **1971**, *75*, 642–648.

111. Katritzky, A. R.; Ignatchenko, E. S.; Barcock, R. A.; Lobanov, V. S.; Karelson, M. *Anal. Chem.* **1994**, *66*, 1799–1807.
112. Murray, J. S.; Politzer, P. In *Theoretical Organic Chemistry*, Parkanyi, C. Ed., Vol. 5, **1998**, pp. 189–202.
113. Haeberlein, M.; Murray, J. S.; Brinck, T.; Politzer, P. *Can. J. Chem.* **1992**, *70*, 2209–2214.
114. Murray, J. S.; Brinck, T.; Politzer, P. *J. Mol. Struct. (Theochem)* **1992**, *255*, 271–281.
115. Brinck, T.; Murray, T.; Politzer, P. *Int. J. Quantum Chem.* **1993**, *48*, 73–88.
116. Shusterman, A. J. *CHEMTECH* **1991**, (Oct.), 624–627.
117. Debnath, A. K.; Lopez de Compadre, R. L.; Debnath, G.; Shusterman, A. J.; Hansch, C. *J. Med Chem.* **1991**, *34*, 786–797.
118. Grigoras, S. *J. Comput. Chem.* **1990**, *11*, 493–510.
119. Stanton, D. T.; Jurs, P. C. *Anal. Chem.* **1990**, *62*, 2323–2329.
120. Egolf, L. M.; Jurs, P. C. *J. Chem. Info. Comput. Sci.* **1993**, *33*, 616–625.
121. Egolf, L. M.; Jurs, P. C. *Anal. Chem.* **1993**, *65*, 3119–3128.
122. Nelson, T. M.; Jurs, P. C. *J. Chem. Info. Comput. Sci.* **1994**, *34*, 601–609.
123. Egolf, L. M.; Wessel, M. D.; Jurs, P. C. *J. Chem. Info. Comput. Sci.* **1994**, *34*, 947–956.
124. Xu, L.; Ball, J. W.; Dixon, S. L.; Jurs, P. C. *Environ. Toxicol. Chem.* **1994**, *13*, 841–851.
125. McGowan, J. C.; Mellors, A. *Molecular Volumes in Chemistry and Biology*; Wiley: New York, **1986**.
126. Bhattacharjee, S. *Computers Chem.* **1993**, *17*, 383–388.
127. Bhattacharjee, S. *Computers Chem.* **1995**, *19*, 51–58.
128. Needham, D. E.; Wei, I. C.; Seybold, P. G. *J. Am. Chem. Soc.* **1988**, *110*, 4186–4194.
129. Karelson, M.; Perkson, A. *Computers Chem.* **1999**, *23*, 49–59.
130. Doucette, W. J.; Andren, A, W. *Environ. Sci. Technol.* **1987**, *21*, 821–824.
131. De Bruijn, J.; Hermens, J. *Quant. ?Struct.-Act. Relat.* **1990**, *9*, 11–21.
132. Pearlman, R. S. In *Physical Chemical Properties of Drugs*, Yalkowski, S. A.; Sinkula, A. A.; Valvani, S. C. Eds., Marcel Dekker: New York, **1980**, Chap. 10, pp. 321–347.
133. Leo, A.; Hansch, C.; Jow, P. Y. C. *J. Med. Chem.* **1976**, *19*, 611–615.
134. Amidon, G. L.; Anik, S. T. *J. Pharm. Sci.* **1976**, *65*, 801–805.
135. Leahy, D. E. *J. Pharm. Sci.* **1986**, *75*, 629–636.
136. Hanai, T.; Hubert, J. *J. Chromatogr.* **1984**, *290*, 197–206.
137. Dunn, W. J. III.; Koehler, M.; Stalling, D. L.; Schwartz, T. R. *Anal. Chem.* **1986**, *58*, 1835–1838.
138. Dunnivant, F. M.; Elzerman, A. W.; Jurs, P. C.; Hasan, M. N. *Environ. Sci. Technol.* **1992**, *26*, 1567–1573.
139. Yalkowsky, S. H.; Valvani, S. C. *J. Chem. Engr. Data* **1979**, *24*, 127.
140. Wiener, H. *J. Am. Chem. Soc.* **1947**, *69*, 17–20.
141. Hosoya, H. *Bull. Chem. Soc. Japan* **1971**, *44*, 2332–2337.
142. Kier, L. B.; Hall, L. H. *Pharm. Res.* **1990**, *7*, 801–807.
143. Hall, L. H.; Kier, L. B. *J. Chem. Info. Comput. Sci.* **1991**, *31*, 76–83.
144. Hall, L. H.; Mohney, B.; Kier, L. B. *Quant. Struct.-Act. Relat.* **1991**, 10, 43–51.
145. Kier, L. B.; Hall, L. H.; Frazer, J. W. *J. Math. Chem.* **1992**, *7*, 229–237.
146. Kier, L. B.; Hall, L. H. *Molecular Structure Description: The Electrotopological State*; Academic Press: San Diego, **1999**.
147. Kier, L. B.; Hall, L. H. *J. Pharm. Sci.* **1981**, *70*, 583–587.
148. Hinze, J.; Jaffe, H. H. *J. Am. Chem. Soc.* **1962**, *84*, 540–549.
149. Joshi, N. Ph.D. thesis, Virginia Commonwealth Univ., Richmond, VA, **1993**.
150. Kier, L. B.; Hall, L. H. *SAR QSAR Environ. Res.* **1995**, *3*, 97–100.

PHASE CHANGE STRUCTURE–PROPERTY QUANTITATIVE RELATIONSHIPS VIA INTERMOLECULAR FORCE PARAMETERS

Marvin Charton

Advances in Quantitative Structure Property Relationships
Volume 3, pages 137–222.

I. INTRODUCTION

Many properties are a function of the difference in intermolecular forces (imf) between compounds in an initial and a final phase. The change in the free energy associated with the contribution of these forces to the property is given by the expression:

$$Q = f(E_{imf_f} - E_{imf_i}) = f(\Delta E_{imf}) \quad (1)$$

Some properties are a function of the difference in intramolecular forces (imf) between compounds in an initial and a final state. This occurs when parts of a molecule are in close proximity to each other. Again, the change in the free energy associated with the contribution of these forces to the property is given by Eq. 1. These properties, and many others, are a function of molecular structure. Such properties can be modeled by parameterizing the forces involved as functions of structural variation and using the resulting correlation equation to obtain a quantitative model of the property of interest by means of multiple linear regression analysis. This work is concerned with obtaining such models.

A. Nature of Structural Effects

1. Structure–Property Quantitative Relationships (SPQR)

In the second half of the nineteenth century the structural theory of organic chemistry was developed. It led to the concept that chemical, physical, and biological properties of all kinds are a function of structural variation. Most of the earliest structure-property relationships (SPR) were qualitative. With the development of methods of quantitative measurement of these properties data accumulated. Attempts were then made to develop quantitative models of the structural dependence of these properties. This was a major interest of early physical chemists.[1–4]

Quantitative descriptions of the structural dependence of properties are called structure property quantitative relationships (SPQR).[5] There are four types of these relationships:

1. Quantitative structure-reactivity relationships (QSRR). Chemical reactivities involve the formation and/or cleavage of chemical bonds. Equilibrium constants, rate constants, polarographic half-wave potentials, and oxidation-reduction potentials are examples of chemical reactivity data.
2. Quantitative structure-chemical property relationships (QSCR). Chemical properties involve a difference in inter- or intramolecular forces between an initial and a final state. Equilibrium constants for hydrogen bonding and charge transfer complex formation, conformational equilibria, partition and distribution coefficients; chromatographic properties such as capacity factors in high performance liquid chromatography, retention times in gas chromatography, and R_F values in thin layer and paper chromatography; melting and boiling points; solvent effects on equilibrium or rate constants; and solubilities are examples of chemical property data.
3. Quantitative structure-physical property relationships (QSPR). Physical properties are either ground or excited state properties, or properties which depend on the difference in energy between the ground state and an excited state. Bond lengths, bond angles, and some dipole moments are ground state or excited state properties, infrared, ultraviolet, nuclear magnetic resonance and other types of spectra, ionization potentials, and electron affinities are properties which depend on the energy difference between states.
4. Quantitative structure-activity relationships (QSAR). Any property associated directly or indirectly with a living organism is a biological activity. The bioactive substrates studied include pure enzymes, tissue homogenates, single-celled organisms, whole tissues, and large multi-

cellular organisms. The data may be obtained in vitro or in vivo. They include rate and equilibrium constants for enzyme reactivity and for binding to receptor sites, various kinds of toxicity determinations such as lethal dose and lethal concentration, and minimum effective concentrations.

a. *The Nature of SPQR*

There are several different types of chemical species, including molecules, ions, radicals, carbenes; nitrenes, benzynes, ...; for which SPQR can be determined. Three kinds of structure are possible:

1. Species with the structure XGY where X is a variable substituent, Y is a constant active site (an atom or group of atoms at which a measurable phenomenon takes place), and G is a constant skeletal group to which X and Y are bonded.
2. Species with the structure XY in which the variable substituent X is directly attached to the constant active site Y.
3. Species in which substituent and active site are the same, the entire species is the active site and it varies. These species are designated X_{Y}.

The purpose of SPQR is to provide a quantitative description of the change in some measurable quantity Q that occurs when a change is made in the structure of the species by varying the substituent X. All of the other pertinent variables such as the conditions of the measurement are held constant. Thus:

$$\left(\frac{\partial Q}{\partial X}\right)_{G,Y,T,P,Sv,I,\ldots} = Q_X \tag{2}$$

where Q_X is the measured quantity when the substituent is X, G is the skeletal group, Y the active site, T the temperature, P the pressure, Sv the solvent, I the ionic strength, and all of these are constant throughout the data set.

We **assume** that Q_X will be a linear function of some number of parameters which represent the effects of the structural variation of X. Then:

$$Q_X = a_1 p_{1X} + a_2 p_{2X} + a_3 p_{3X} + \cdots + a_0 \tag{3a}$$

$$= \sum_{i=1}^{n} a_i p_{iX} + a_0 \tag{3b}$$

where the p_i are the parameters which account for the structural effect of X on Q. These parameters have been obtained in various ways:

1. From quantum chemical calculations.[6] This method is most suitable for electrical effect parameters.
2. From molecular mechanics calculations[7] for steric effect parameters.
3. From a reference set by definition (primary values). This method assumes that structural effects on the data set to be studied are a linear function of those which occur in the reference set. Secondary values of these parameters can be estimated by various methods. This is the method generally used to parameterize QSRR.
4. From comparative molecular field analysis (COMFA).[8] This method can be used for electrical, steric, and polarizability parameters.
5. From molecular geometry for steric parameters.[9–11]
6. From topological methods. This method is best restricted to steric effect and polarizability parameters. Electrotopological parameters seem to be a function of polarizability and electronegativity.[12–14]

When suitable parameters are available the values of Q can be correlated with them by means of either simple linear regression analysis if the model requires only a single variable or multiple linear regression analysis if it requires two or more variables. Such a correlation results in a SPQR. In this work we consider only those parameters that are defined directly or indirectly from suitable reference sets or in the case of steric parameters calculated from molecular geometries.

b. The Uses of SPQR

SPQR have three major uses:

1. Mechanistic. QSRR and those QSAR which involve enzyme reactivity can provide information about the sensitivity of a reaction to electrical effects, its electronic demand, the composition of the electrical effect, and the sensitivity to steric effects. QSAR which involve binding to receptor sites can provide information about the nature of the receptor site. Other QSAR can shed light on the bioactivity determining step.
2. Predictive: All SPQR can be used to predict reactivities, chemical and physical properties, and bioactivities. There are manifold practical applications of such predictions. Particular examples include the design of bioactive molecules such as medicinal drugs and pesticides. In addition to the maximization of activity and minimization of side effects desirable pharmaceutical properties such as improved solubility, longer shelf life, and controlled release can be developed. They are a major method in environmental science where they can be used to predict toxicities, biodegradabilities, and other properties of environmental interest. Finally, they are of use in materials science in the development of materials with specific properties.

3. Archival. SPQR provide a concise, efficient, and convenient method for storing the results of experimental studies on the effect of structural changes upon properties of all kinds.

c. *The Types of Structural Effects*

Structural effects are conveniently divided into three categories:

1. Electrical effects. These effects cause a variation in the electron density at the active site. They account for the ability of a substituent to stabilize or destabilize a cation, anion, radical, carbene, or other chemical species.
2. Steric effects. These effects result from the repulsion between valence electrons in orbitals on atoms which are in close proximity but not bonded to each other.
3. Inter- and intramolecular force effects. These effects result either from the interactions between the substituent and its immediate surroundings such as the medium, a surface, or a receptor site or from the effect of the substituent on the interactions of the skeletal group G and the active site Y with their surroundings.

Electrical effects are the major factor in chemical reactivities and physical properties. Intermolecular forces are usually the major factor in bioactivities. Either electrical effects or intermolecular forces may be the predominant factor in chemical properties. Steric effects only occur when the substituent and the active site are in close proximity to each other and even then rarely account for more than twenty five percent of the overall substituent effect.

B. Electrical Effects

Electrical effects are conveniently described by the triparametric (three independent variables) LDR equation:[15]

$$Q_X = L\sigma_{lX} + D\sigma_{dX} + R\sigma_{eX} + h \quad (4)$$

or relationships derived from it. The parameters are:

σ_l, the localized (field) electrical effect parameter. It is identical to σ_I and σ_F. Though other localized electrical effect parameters such as $\sigma_l{}^q$ have been proposed there is no advantage to their use. The σ^* parameter and the F parameter have sometimes been used as localized electrical effect parameters, such use is generally incorrect as both of these parameters contain a small but significant delocalized effect contribution. As was noted above the available evidence is strongly in favor of an electric field model for transmission of the localized effect.

σ_d, the intrinsic delocalized (resonance) electrical effect parameter. It represents the delocalized electrical effect in a system with no electronic demand.

σ_e, the electronic demand sensitivity parameter. It adjusts the delocalized effect of a group to meet the electronic demand of the system.

The electrical effect is characterized by two quantities derived from Eq. 1:

The electronic demand, η, is a property of a system or of a composite electrical effect parameter that is itself a function of both σ_d and σ_e. It is defined as R/D where R and D are the coefficients of σ_e and σ_d respectively.

The per cent delocalized effect, P_D. It is defined as:

$$P_D = \frac{100D}{L + D} \tag{5}$$

C. Steric Effects

1. Introduction

Steric effects[9–11] can be of several types. What they have in common is that their magnitude is related to their size.

a. Primary Steric Effects

These effects are due to repulsions between electrons in valence orbitals on adjacent atoms which are not bonded to each other. They supposedly result from the interpenetration of occupied orbitals on one atom by electrons on the other resulting in a violation of the Pauli exclusion principle. **All primary steric interactions raise the energy of the system in which they occur**. Their effect on chemical reactivity is to either decrease or increase a rate or equilibrium constant depending on whether steric repulsions are greater in the reactant or in the product (equilibria) or transition state (rate).

b. Secondary Steric Effects

These are effects on chemical reactivity that may result from the shielding of an active site from the attack of a reagent, from solvation, or both. They may also be due to a steric effect that determines the concentration of the reacting conformation of a chemical species. The secondary steric effect of a nonsymmetric group will also depend on its conformation.

c. Direct Steric Effects

These effects can occur when the active site at which a measurable phenomenon occurs is in close proximity to the substituent. Among the many skeletal groups exhibiting direct steric effects are vicinally (1,2) substituted skeletal groups such as *ortho* substituted benzenes, **1**, *cis*-substituted ethylenes, **2**, the *ortho* (1,2-; 2,1-; and 2,3-) naphthalenes, **3**, **4**, and **5** respectively; and *peri* (1,3) skeletal groups such as 1,8-substituted naphthalenes, **6**. Other vicinal examples are *cis*-1,2- disubstituted. cyclopropanes, *cis*-2,3-disubstituted norbornanes, and *cis*-2,3-disubstituted [2.2.2]-bicyclooctanes, **7**, **8**, and **9** respectively. Some skeletal groups do not always show steric effects. 2,3-disubstituted five membered ring heteroarenes such as thiophenes and selenophenes are generally free of

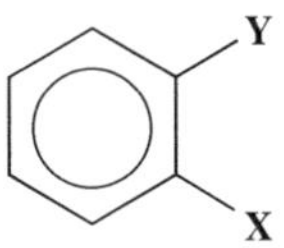

1

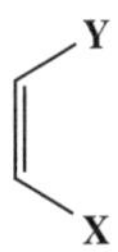

2

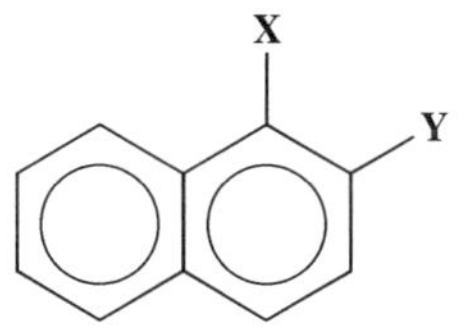

3

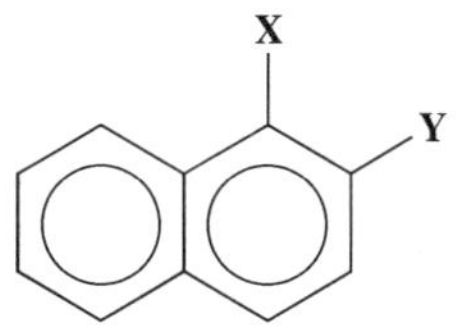

4

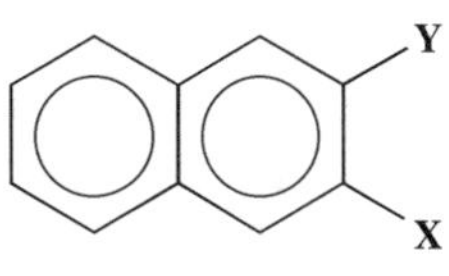

5

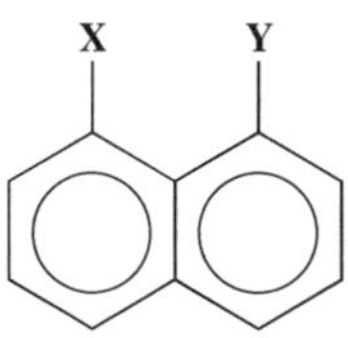

6

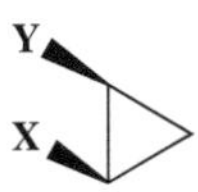

7

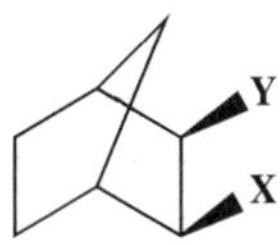

8

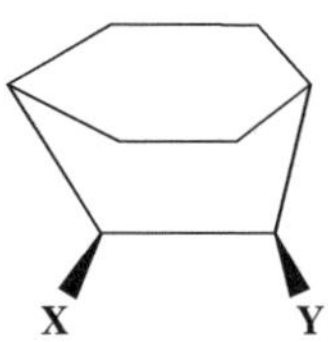

9

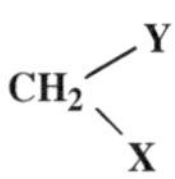

10

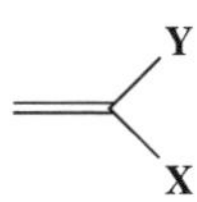

11

steric effects. This is probably due to the larger XCC angle in these systems as compared with ethene and benzene systems. Geminally substituted (1,1) skeletal groups such as disubstituted methanes, **10**, and 1,1-disubstituted ethenes, **11**, are also generally free of steric effects.

d. Indirect Steric Effects

These effects are observed when the steric effect of the variable substituent is relayed by a constant substituent between it and the active site as in **12** where Y is the active site, Z is the constant substituent, and X is the variable substituent. This is a type of buttressing effect.

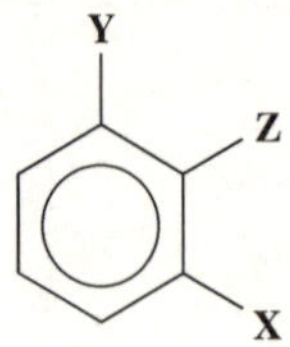

12

e. Steric Effects Generally are Vector Quantities

This is easily shown by considering for example the ratio r of the steric parameter υ for any five carbon alkyl group to that for 1-pentyl. Values of r are: 1-Pe, 1; 2-Pe, 1.54; 3-Pe, 2.22; CH_2sBu, 1.47; CH_2iBu; 1.00; CH_2tBu, 1.97; CMe_2Pr, 2.40; CHiPrMe, 1.90. All of these groups have the same volume and therefore the same bulk but they differ in steric effect.[16] In order to account for this it is necessary to consider what happens when a non-symmetric substituent is in contact with an active site. We consider the simple case of a spherical active site Y in contact with a nonsymmetric substituent, $CZ^LZ^MZ^S$, where the superscripts, L, M, and S represent the largest, the medium sized and the smallest Z groups respectively. The C—G bond and the Y—G bond are of comparable length. There are three possible conformations of this system (Figure 1). As all steric repulsions raise the energy of the system, the preferred conformation will be the one that results in the lowest energy increase. This is the conformation which presents the smallest face to the active site, conformation **C**. From this observation we have the minimum steric interaction (MSI) principle which states: **a non-**

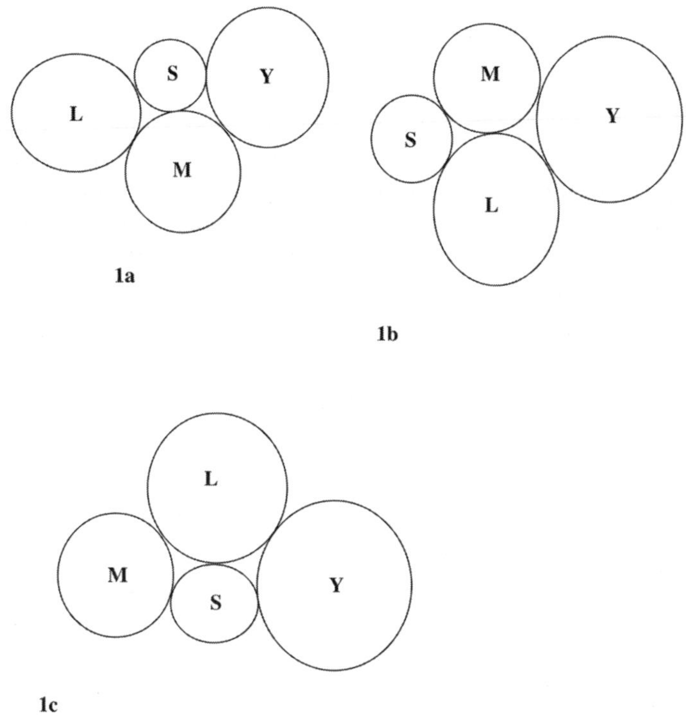

Figure 1. Top views of possible conformations of a molecule with a spherical active site Y adjacent to a tetrahedral substituent $MZ^LZ^MZ^S$ where L, M, and S designate the largest, medium-sized, and smallest groups, respectively. Conformation A has the lowest energy, conformation C has the highest.

symmetric substituent prefers that conformation which minimizes steric interactions. The directed nature of steric effects results in a conclusion of vital importance: that in general **the volume of a substituent is not an acceptable measure of its steric effect**.[16] There are still some workers who are unable to grasp this point. It is nevertheless true that group volumes are not useful as steric parameters except in the case of substituents that are roughly spherical and not always then. They are actually measures of group polarizability. In short, for a range of different substituent shapes in a data set **steric effects are not directly related to bulk, polarizability is**.

2. *The Monoparametric Model of Steric Effects*

a. *Introduction*

Van der Waals radii, r_v, have long been held to be an effective measure of atomic size.[17] Charton proposed the use of the van der Waals radius as a steric parameter[18] and developed a method for the calculation of group van der Waals radii for tetracoordinate symmetric top substituents MZ_3 such as the methyl and trifluoromethyl groups.[19] In later work the hydrogen atom was chosen as the reference substituent and the steric parameter υ was defined as

$$\upsilon_X \equiv r_{VX} - r_{VH} = r_{VX} - 1.20 \qquad (6)$$

where r_{VX} and r_{VH} are the van der Waals radii of the X and H groups in Angstrom units.[20] Expressing r_V in these units is preferable to the use of picometers because the coefficient of the steric parameter is then comparable in magnitude to the coefficients of the electrical effect parameters. Whenever possible, υ parameters are obtained directly from van der Waals radii or calculated from them. An equation has been derived which makes possible the calculation of υ values for nonsymmetric tetrahedral groups of the types $MZ_2^SZ^L$ and $MZ^SZ^MZ^L$ in which the Z groups are symmetric.[21] These are considered to be primary values. For the greater number of substituents however, υ parameters must be calculated from the regression equations obtained for correlations of rate constants with primary values. The values obtained in this manner are considered to be secondary υ values. All other measures of atomic size are a linear function of van der Waals radii.[10] There is, therefore, no reason for preferring one measure of atomic size over another. As values of υ were developed for a wide range of substituent types with central atoms including oxygen, nitrogen, sulfur, and phosphorus as well as carbon, these parameters provide the widest structural range of substituents for which a measure of the steric effect is available.

b. *Steric Classification of Substituents*

This is useful in understanding the way in which different types exert steric effects.[9,10] Substituents may be divided into three categories based on the degree of conformational dependence of their steric effects:

1. No conformational dependence (NCD). Groups of this type include monatomic substituents such as hydrogen and the halogens; $M^a \equiv M^b$ substituents such ethynyl and cyano, and MZ_3 groups.
2. Minimal conformational dependence (MCD). Among these groups are:

(a) Nonsymmetric substituents with the structure $MH_n(lp)_{3-n}$ such as the hydroxyl and amino groups (lp is a lone pair)
(b) Nonsymmetric substituents with the structure $MZ_2^SX^L$ where S stands for small and L for large.

3. Strong conformational dependence (SCD). These groups have the structures:

(a) $MZ_2^LZ^S$ and $MZ^LZ^MZ^S$, where the superscript M indicates medium.
(b) Planar π-bonded groups MZ^LZ^S where M and either or both Z's are sp^2 hybridized, such as phenyl, acetyl, nitro ($X_{p\pi}$ groups).
(c) Quasi planar π-bonded groups such as dimethylamino and cyclopropyl.

The steric parameter for NCD groups can be obtained directly from van der Waals radii or calculated from them. The values for MCD groups are often obtainable from van der Waals radii although in some cases they must be derived as secondary values from regression equations obtained by correlating rate constants with known values of the steric parameter. Steric parameters for SCD groups of the nonsymmetric type are usually obtainable only from regression equations. In the case of planar π-bonded groups the maximum and minimum values of the steric parameter are available from the van der Waals radii. For a more detailed discussion see ref. 1.

3. *Multiparametric Models of Steric Effects*

a. *Introduction*

In some cases a simple monoparametric model of the steric effect is insufficient. Examples are when the active site is itself large and nonsymmetric or alternatively when the phenomenon studied is some form of bioactivity in which binding to a receptor determines activity. The failure of the monoparametric model is due to the fact that a single steric parameter cannot account for the variation of the steric effect at various points in the substituent. The use of a multiparametric model of steric effects that can represent the steric effect at different segments of the substituent is required. Five multiparametric models are available; that of Verloop,[22] the simple branching model, the expanded branching model, the segmental model, and the composite model. The Verloop model will not be discussed as it is flawed.

b. *Branching Models*

The simple branching model[10] for the steric effect is given by the expression

$$S\psi = \sum_{i=1}^{m} a_i n_i + a_b n_b \tag{7}$$

where $S\psi$ represents the steric effect parameterization, the a_i and a_b are coefficients, n_i is the number of branches attached to the ith atom, and n_b is the number of bonds between the first and last atoms of the group skeleton. It follows that n_b is a measure of group length. Unfortunately it is frequently highly collinear in group polarizability, which greatly limits its utility. For saturated cyclic substituents it is necessary to determine values of n_i from an appropriate regression equation. For planar π-bonded groups n_i is taken to be 1 for each atom in the group skeleton. For other groups n_i is obtained simply by counting branches. The model makes the assumption that all of the branches attached to a skeleton atom are equivalent. This is at best only a rough approximation. Distinguishing between branches results in an improved model called the expanded branching equation:

$$S\psi = \sum_{i=1}^{m} \sum_{j=1}^{3} a_{ij} n_{ij} + a_b n_b \tag{8}$$

which allows for the difference in steric effect that results from the order of branching.[10] This difference follows from the MSI principle. The first branch has the smallest steric effect because a conformation in which it is rotated out of the way of the active site is preferred. In this conformation the active site is in contact with two hydrogen atoms. The preferred conformation in a the case of a second branch has the larger of the two branches directed out of the way. The smaller branch and a hydrogen atom are in contact with the active site. When there are three branches the largest will be directed out of the way and the other two will be in contact with the active site. The problem with the expanded branching method is that it requires a large number of parameters. Data sets large enough to permit its use are seldom seen.

c. *The Segmental Model*[10]

The segmental model is often the simplest and most effective of the multiparametric models. In this model each atom of the group skeleton together with the atoms attached to it constitutes a segment of the substituent. Applying the MSI principle, the segment is considered to have that conformation which presents its smallest face to the active site. The segment is assigned the υ value of the group which it most resembles. Values of the segmental steric parameters υ_i, where i designates the segment number are given in Table 8. Numbering starts from the first atom of the group skeleton which is the atom that is attached to the rest of the system. The segmental model is given by expression:

$$S\psi = \sum_{i=1}^{m} S_i \upsilon_i \tag{9}$$

d. *The Composite Model*

This is a combination of the monoparametric υ model with the simple branching model. This method has proven useful in modeling amino acid, peptide, and protein properties.[10] It is an improvement over the simple branching model and requires only one additional parameter.

D. Intermolecular Forces

1. Properties Dependent on Intermolecular Forces

Properties dependent largely or entirely on intermolecular forces are given in Table 1. They include phase change, phase transfer, solution, complex formation, conformation and bioactivity.

2. Types of Intermolecular Forces

The types of intermolecular forces most commonly encountered are summarized in Table 2 with the variables upon which they depend.

3. Parameterization of Intermolecular Forces

a. *Hydrogen Bonding*

Hydrogen bonding requires two parameters for its description, one to account for the hydrogen atom donating capacity of a substituent and another to account for its hydrogen atom accepting capacity. A simple approach is the use of n_H, the number of OH and/or NH bonds in the substituent and n_n, the number of lone pairs on oxygen and/or nitrogen atoms as parameters.[23,24] The use of these parameters is based on the argument that if one of the phases involved in the phenomenon studied includes a protonic solvent, particularly water, then all of the hydrogen bonds the substituent is capable of forming will indeed form. For such a

Table 1. Quantities that depend on ΔE_{imf}

Property type	*Q*	*Property*	φ_i	φ_r
Phase change	T_m	Melting point	(s)	(l)
	T_b	Boiling point	(l)	(g)
	$(\Delta U_V^o)_o$	Cohesive energy (0 °C)	(l)	(g)
Phase transfer	E_C	Cohesive energy, polymers		
	log *Sol*	Solubility	(s) or (1)	Solution
	log *P*	Partition coefficient	$Sv_1(Sv_2)$	$Sv_2(Sv_1)$
	log *D*	Distribution coefficient	$Sv_1(Sv_2)$	$Sv_2(Sv_1)$
	log t_w	GC	(g)	Column
	log *K′*	HPLC	Solution	Column
	log R_F	TLC, RPTLC, Paper	Solution	Surface
	ΔG_{tr}	Transfer	Sv_1	Sv_2
	ΔG_{st}	Surface tension		
Solvent effect	log *k*	log rate constant		
Complex formation	log *K*	Complex formation, pmr	Pmr + XGY	[Pmr–XGY]
	log K	Charge Transfer Complex	XGY + Z	[XGY–Z]
Conformation	T_g	Glass temperature	$Confm_1$	$Confm_2$
	P_{Xaa}	Conmf. Probability, Protein		
	$\Delta G_{h/c}$	Helix-Coil, Protein	Coil	Helix
	ΔG_{tr}	Globulin interior to surface		
Bioactivity	$LD_{\%}$	Lethal dose	Solution	Receptor
	MEC	Minimum effective conc.	Solution	Receptor
	$LC_{\%}$	Lethal conc.	Air	Receptor
	MIC	Minimum inhibitory conc.	Solution	Receptor

Abbr.: Sv, solvent; pmr, polymer; Confm, conformation.

system, hydrogen bond parameters defined from equilibria in highly dilute solution in an "inert" solvent are unlikely to be a suitable model. This parameterization accounts only for the number of hydrogen donor and hydrogen acceptor sites in a group. It does not take into account differences in hydrogen bond energy. A more effective parameterization than that

Table 2. Intermolecular forces and the quantities upon which they depend.[5]

Intermolecular force	*Quantity*
molecule–molecule	
Hydrogen bonding (hb)	q_{M^H}, q_{M^E}, orbital type
Dipole–dipole (dd)	dipole moment
Dipole–induced dipole (di)	dipole moment, polarizability
Induced dipole–induced dipole (ii)	polarizability
Charge transfer (ct)	ionization potential, electron affinity
ion–molecule	
ion–dipole (Id)	ionic charge, dipole moment
ion–induced dipole (Ii)	ionic charge, polarizability

Abbreviations are in parentheses. dd interactions are also known as Keesom interactions; di interactions are also known as Debye interactions; ii interactions are also known as London or dispersion interactions. Collectively, dd, di, and ii interactions are known as Van der Waals interactions. Charge transfer interactions are also known as donor–acceptor interactions.

above would be the use of the hydrogen bond energy for each type of hydrogen bond formed.[25] Then for each substituent the parameter E_{hbX}, would be given by the equation

$$E_{hbX} = \sum_{i=1}^{m} n_{hbi} E_{hbi} \tag{10}$$

where E_{hbX} is the hydrogen bonding parameter, E_{hbi} is the energy of the ith type of hydrogen bond formed by the substituent X, and n_{hbi} is the number of such hydrogen bonds. The validity of this parameterization remains to be determined. The site number parameterization suffers from the fact that though it accounts for the number of hydrogen bonds formed it does not differentiate between their energies and can therefore be only an approximation. A recent definition of a scale of hydrogen bond acceptor values from 1-octanol-water partition coefficients of substituted alkanes shows that the site number method strongly overestimates the hydrogen acceptor capability of the nitro group and seriously underestimates that of the methylsulfoxy group.[26] Much remains to be done in properly parameterizing hydrogen bonding.

Hydrogen bonding may also involve the effect of the substituent on some constant group (present in each member of the data set) which can act as either a hydrogen bond donor or a hydrogen bond acceptor. In that case the magnitude of the hydrogen bonding is dependent on the electrical effect of the substituent x and if the group is in close proximity to the substituent on steric effects of X as well.

b. van der Waals Interactions

These interactions (dd, di, ii) are a function of dipole moment and polarizability. It has been shown that the dipole moment cannot be replaced entirely by the use of electrical effect substituent constants as parameters.[27,28] This is because the dipole moment has no sign. Either an overall electron donor group or an overall electron acceptor group may have the same value of μ. It has also been shown that the bond moment rather than the molecular dipole moment is the parameter of choice. The dipole moments of MeX and PhX were taken as measures of the bond moments of substituents bonded to sp^3 and sp^2 hybridized carbon atoms respectively of a skeletal group. Application to substituents bonded to sp hybridized carbon atoms should require a set of dipole moments for substituted ethynes. Dipole moments for MeX and PhX can be calculated from the equations:[27]

$$\mu_{MeX} = 5.11(\pm 0.497)\sigma_{lX} + 1.99(\pm 0.541)\sigma_{dX} + 0.0129(\pm 0.205) \quad (11)$$

and:

$$\mu_{PhX} = 5.47(\pm 0.150)\sigma_{lX} + 4.30(\pm 0.446)\sigma_{dX} + 6.94(\pm 1.91)\sigma_{eX} + 0.420(\pm 0.172) \quad (12)$$

The polarizability parameter used in most of this work, α, is given by the expression

$$\alpha \equiv \frac{MR_X - MR_H}{100} = \frac{MR_X}{100} - 0.0103 \quad (13)$$

where MR_X and MR_H are the group molar refractivities of X and H respectively.[23,24,29] The factor 1/100 is introduced to scale the α parameter so that its coefficients in the regression equation are roughly comparable to those obtained for the other parameters used. If all of the carbon atoms in a compound have the same hybridization then n_C, the number of these atoms, can be used as a polarizability parameter. There are many other polarizability parameters including parachor, group molar volumes of various kinds, van der Waals volumes, and accessible surface areas any of which will do as well as they are all highly collinear in each other.[30] Proposing other polarizability parameters has been a cottage industry in the past.

Values of α can be estimated by additivity from the values for fragments or from group molar refractivities calculated from the equation

$$MR_X = 0.320n_c + 0.682n_b - 0.0825n_n + 0.991 \quad (14)$$

where n_c, n_b, and n_n are the number of core, bonding, and nonbonding electrons respectively in the group X.[30]

Atomic polarizabilities P_M for atoms M have been collected by Miller.[31] They may be used as polarizability parameters but they are not directly comparable to α values, thus a comparison of coefficients obtained from correlations with α are not directly comparable to those obtained with P_M. in order to obtain α values for some atoms we have first defined:

$$\Delta P_M \equiv P_M - P_H \quad (15)$$

where M is the atom of interest and H is hydrogen. This is directly comparable to the definition of α in Eq. 11. ΔP_M values for the set H, F, Cl, Br, I (Table 3) were then correlated with the equation:

$$\Delta P_M = A\alpha_M + a_o \quad (16)$$

to give the regression equation:

$$\Delta P_{at,M} = 31.4(\pm 0.428)\alpha - 0.0507(\pm 0.0305) \quad (17)$$

$100r^2$, 99.94; F, 5396; S_{est}, 0.0472; S°, 0.0304; N_{dp}, 5.

Inclusion in the data set of the $\Delta P_{at,M}$ values for N, P, O, S, and Se (Table 3) gave the regression equation:

$$\Delta P_M = 30.6(\pm 1.70)\alpha_M - 0.197(\pm 0.126) \quad (18)$$

$100r^2$, 97.26; F, 320.0; S_{est}, 0.248; S°, 0.183; N_{dp}, 11.

Though the goodness of fit of Eq. 18 is decidedly poorer than that of Eq. 17, the values of A are not significantly different and the intercepts in both equations are not significant, furthermore, both equations are significant at the 99.9% confidence level. $\Delta P_{at,M}$ values for the Group 14 elements seem to

Table 3. ΔP_M and α values of the atoms M

M, ΔP, α: H, 0, 0; F, −0.11, −0.001; Cl, 1.513, 0.0510; Br, 2.383, 0.079; I, 4.033, 0.129; O, 0.135, 0.015; S, 2.233, 0.082; Se, 3.103, 0.114; N, 0.433, 0.038; P, 2.963, 0.107; C, 1.093, 0.039; He, −0.462, −0.017; Ne, −0.271, −0.010; Ar, 0.974, 0.029; Kr, 1.818, 0.056; Xe, 3.377, 0.106; Rn, 4.633, 0.146.

α values for the rare gases are calculated from equation 19.

lie on their own line. We have used Eq. 17 in the form:

$$\alpha_M = 0.0318\Delta P_M - 0.00183 \quad (19)$$

to calculate α values for the rare gas elements. These values are also given in Table 3.

c. Charge Transfer Interactions

These interactions can be roughly parameterized by the indicator variables n_{A} and n_{D}. n_{A} takes the value 1 when the substituent is a charge transfer acceptor and 0 when it is not. n_{D} takes the value 1 when the substituent is a charge transfer donor and 0 when it is not. An alternative parameterization makes use of the first ionization potential of MeX (ip_{MeX}) as the electron donor parameter and the electron affinity of MeX as the electron acceptor parameter.[25] Usually, the indicator variables n_{A} and n_{D} are sufficient. This parameterization accounts for charge transfer interactions directly involving the substituent. If the substituent is attached to a π-bonded skeletal group, then the skeletal group is capable of charge transfer interaction the extent of which is modified by the substituent. This is accounted for by the electrical effect parameters of the substituent.

d. The Intermolecular Force (IMF) Equation

A general relationship for the quantitative description of intermolecular forces called the intermolecular force (IMF) equation is:

$$Q_X = L\sigma_{lX} + D\sigma_{dX} + R\sigma_{eX} + M\mu_X + A\alpha_X + H_1 n_{HX} + H_2 n_{nX} + Ii_X + B_{DX}n_{DX} + B_{AX}n_{AX} + S\psi_X + B_o \quad (20)$$

Variations of this equation will be used throughout this work.

II. METHOD

A. Statistics

In all of this work data sets have been correlated with equations using simple linear regression analysis in the case of a single independent variable and multiple linear regression analysis in the case of two or more independent variables. The statistics reported for multiple linear regression analysis include the $100R^2$ statistic that represents the per cent of the variance of the data accounted for by the regression equation. The $\mathrm{A}100R^2$

statistic is the adjusted value of $100R^2$, it takes into account the number of independent variables in the regression equation. The F statistic is a measure of the significance of the correlation, its confidence level is given in parentheses. The S_{est} statistic is the standard error of the estimate, it is useful in detecting outliers. It is of no use as a measure of the goodness of fit of different data sets to a model if they involve different types of data as it depends on the nature of the data correlated. The S^o statistic may be thought of as a standard error of the estimate which has been corrected for the nature of the data, it is a useful measure of the goodness of fit of the data to the regression equation. With the exception of $100R^2$ which is replaced by $100r^2$ the same statistics are used for simple linear regression analysis. The regression equation reported is that which results after the elimination of variables whose coefficients are not significant. Statistics are reported beneath the regression equation.

B. The Quantitative Description of Dependent Variable Composition

It is helpful in discussing results to use C_i, the percent contribution of each independent variable in the regression equation. This quantity is given by:[25]

$$C_i = \frac{100a_ix_i}{\sum_1^m a_ix_i} \qquad (21)$$

where a_i is the regression coefficient of the ith independent variable and x_i is its value for some reference group. We can define a hypothetical reference group or choose an actual reference group. The C_i values are particularly useful in comparing the relative importance of the contribution of the different independent variables to the regression equation both within a given data set and between data sets.

III. PHASE CHANGE PROPERTIES

A. Introduction

In each of the three types of phase change property we begin by considering the simplest possible case and applying the appropriate version of the IMF equation to it. We then proceed through more complicated examples until we reach the most complex that we have studied so far. Unless otherwise noted all data used in this work is taken from refs. 32 to 35.

B. Boiling Point, T_b

1. Monatomic Species

The boiling points of the rare gases, He, Ne, Ar, Kr, Xe, and Rn; Set TB1, Table 4. All of the parameter values required for the correlations are given in Table 5. As these monatomic species have no dipole moment only induced dipole—induced dipole forces need be considered and Eq. 20 simplifies to:

$$Q = A\alpha + a_o \tag{22}$$

On correlation with Eq. 22 the regression Eq. 23 is obtained. The polarizability parameters used are given in Table 5.

$$T_{b,M} = 1219(\pm 82.1)\alpha - 39.5(\pm 6.44) \tag{23}$$

$100r^2$, 98.22; F, 220.2; S_{est}, 11.9; S^{o}, 0.164; N_{dp}, 6.

Clearly, the model accounts vary well for the boiling points of the rare gases. They are dependent only on polarizability.

2. Linear Diatomic Molecules

Almost as simple a case are the boiling points of homonuclear diatomic molecules, M_2 (set TB2a, Table 4). As these molecules also have a dipole moment of zero the correlation equation is again Eq. 22. The regression equation obtained is:

$$T_{b,M_2} = 1643(\pm 123)\alpha + 51.3(\pm 15.0) \tag{24}$$

$100r^2$, 97.26; F, 177.2; S_{est}, 29.3; S^{o}, 0.196; N_{dp}, 7.

Inclusion of heteronuclear diatomic molecules other than hydrogen halides in the data set (set TB2b, Table 4) requires the addition of a term in the dipole moment to the correlation equation which then becomes:

$$Q = A\alpha + M\mu + a_o \tag{25}$$

Parameter values are set forth in Table 5. The regression equation obtained for set TB2b is:

$$T_{b,M^1M^2} = 1578(\pm 93.4)\alpha + 36.4(\pm 16.1)\mu + 55.3(\pm 11.3) \tag{26}$$

$100R^2$, 96.87; A$100R^2$, 96.61; F, 170.2; S_{est}, 26.7; S^{o}, 0.200; N_{dp}, 14; $r_{\alpha\mu}$, 0.287; C_α, 89.7; C_μ, 10.3.

The C_i values are based on the hypothetical reference molecule for which $\alpha = 0.2, \mu = 1$. Polarizability is the major factor in determining boiling point

Table 4. Boiling points used in correlations of sets TB1 to TB9 and cohesive energies of set CE1

Set TB1. Rare gases. M, T_b: He, 4.215; Ne, 27.05; Ar, 87.28; Kr, 119.75; Xe, 165.05; Rn, 211.15.
Set TB2. Diatomic molecules. M^1M^2, T_b: H_2, 20.28; F_2, 85.01; Cl_2, 238.55; Br_2, 331.93; I_2, 457.5; N_2, 77.15; O_2, 90.11; CO, 81.65; NO, 121.35; ICl, 370.55; IBr, 389.15d; ClF, 172.35; BrCl, *278.15*; BrF, 253.15; HCl, 188.25; HBr, 206.15; HI, 237.8.
Set TB3a. Linear triatomic M^1CM^2, T_b: SCS, 319.41; SeCSe, 398.25; OCS, 222.92; SCSe, 357.65; OCSe, 251.45.
Set TB3b. Linear triatomic M^1ZM^2, T_b: FCN, 227.15; ClCN, 286.15; BrCN, 334.5; MeCN, 354.75; MeNC, 332.75.
Set TB4a. Trigonal planar M_3B. M, T_b: F, 172.15; Cl, 285.65, Br, 364.45; I, 483.15; Me, 293.15.
Set TB4b. Trigonal planar $M^1M^2C{=}Z^a$. M^1, M^2, T_b: F, F, 190.05; Cl, Cl, 281.45; Br, Br, 337.65; F, H, 244.15; Me, Me, 329.35, Me, H, 293.95; H, H, 252.15[b].
Set TB5. T_b, °K, $MX^1X^2X^3X^4$. $MX^1X^2X^3X^4$, T_b: SiH_4, 161; SiF_4, 187; $SiCl_4$, 331; $SiBr_4$, 427; SiI_4, 576; $SiMe_4$, 300; GeH_4, 185; $GeCl_4$, 356; $GeBr_4$, 460; GeI_4, 621; $GeMe_4$, 317; SnH_4, 221; $SnCl_4$, 388; $SnBr_4$, 480; SnI_4, 638; $SnMe_4$, 351; $PbMe_4$, 383; SiH_3Br, 275; SiH_3Cl, 243; SiH_3I, 319; SiH_3Me, 216; $SiCl_3Me$, 340; SiF_3H, 368; SiF_3Cl, 203; SiH_2Cl_2, 281; $SiCl_3H$, 306; $SiBr_3H$, 382; $SiMe_2H_2$, 253; GeH_3Br, 325; GeH_3Cl, 301; GeH_2Cl_2, 343.
Set TB6. MH_n. T_b, M, *n*: C, 4, 112; Si, 4, 161; Ge, 4, 185; Sn, 4, 221; *N*, 3, 240; P, 3, 185; As, 3, 211; Sb, 3, 255; Bi, 3, 295; *O*, 2, 373; S, 2, 213; Se, 2, 231; Te, 2, 271; *F*, 1, 293; Cl, 1, 188; Br, 1, 206; I, 1, 238.
Set TB7. MMe_n. T_b, M, *n*: C, 4, 283; Si, 4, 300; Ge, 4, 317; Sn, 4, 351; Pb, 4, 383; *N*, 3, 276; P, 3, 311; As, 3, 324; Sb, 3, 354; Bi, 3, 383; *O*, 2, 248; S, 2, 310; Se, 2, 328; Te, 2, 365; *F*, 1, 195; Cl, 1, 250; Br, 1, 277; I, 1, 316.
Set TB8. 2-substituted-1,3-butadienes[5]. X, T_b: H, 268.75; Me, 307.15; OMe, 348.15; OEt, 368.15; Cl, 332.55; F, 285.15; I, 385.15.
Set TB9. 1-substituted-1,3-butadienes,[5] 1-substituted-4-methyl-1,3-butadienes, X, T_b, T_b: H, 268.75, 315.15; Me, 315.15, 355.15; Et, 346.15, 381.15; OMe, 364.65, —; OEt, 383.15, —; Cl, 341.15, —; CHO, —, 446.65, —; CO_2Me,—, 453.15; CO_2Et, —, 468.15; CN, 409.65,—; CH=CHMe, —, 420.65; Vi, 371.65, —; C_2H, 356.55, —.
Set TB10. Substituted 1,2-propadienes.[5] X, T_b: H, 238.65; Me, 283.95; Et, 318.05; Pr, 249.15; CH_2Br, 383.15; CH_2Cl, 361.15; CH_2I, 403.15; CH_2OMe, 361.15.
Set CE1. MeX. X, (ΔU^o_V) (kJ/mol) at 298.15° K.[38] Me, 7.67; Et, 13.86; tBu, 19.94; Vi, 13.66; $ViCH_2$, 13.66; E—CH=CHMe, 19.51; $H_2C{=}C{=}CH$, 21.21; Ph, 35.58; CH_2Ph, 39.79; 2-Tpn, 36.43; 3-Tpn, 36.98; NO_2, 35.88; C_6F_5, 38.66; CH_2CN, 33.71; CCl_3, 30.14; $CONMe_2$, 47.76; $CHCl_2$, 28.30; CH_2Br, 25.78; CH_2I, 29.57; NMe_2, 19.72; CN, 30.92; OMe, 16.87; OPh, 44.43; CHO, 23.65; Ac, 28.79; COEt, 32.44; OAc, 30.02; CO_2Me, 30.02; OBz, 53.09; SMe, 25.18; SSMe, 35.42; SEt, 29.52, Vi, 13.66.

Note: T_b values are from ref. 34–36 unless otherwise noted. All melting and boiling points and critical temperatures used in correlations in this work are in degrees Kelvin. d., decomposes.

Table 5. Parameter values for sets TB1 to TB9, TM1 to TM5, and TC1 to TC3

Sets TB1, TM1, TC1. M, α; He, −0.017; Ne, −0.010; Ar, 0.029; Kr, 0.056; Xe, 0.106; Rn, 0.146.
Sets TB2, TM2, TC2. M^1M^2, α, μ, ω; F_2, −0.002, 0, 1; Cl_2, 0.100, 0, 1; Br_2, 0.158, 0, 1; I_2, 0.258, 0, 1, H_2, 0, 0, 1; N_2, 0.031, 0, 1; O_2, 0.028, 0, 1; CO, 0.031, 0.128, 0.5; NO, 0.030, 0.158, 0.5; ICl, 0.179, 0.65, 0.5; IBr, 0.208, 1.21, 0.5; ClF, 0.049, 0.881, 0.5; BrCl, 0.129, 0.57, 0.5; BrF, 0.078, 1.29, 0.5; Hcl, 0.060, 1.0929, 0.5; HBr, 0.089, 0.826, 0.5; HI, 0.139, 0.4475, 0.5.
Sets TB3a, TM3a. M^1CM^2, α, μ, ω; SCS, 0.233, 0, 1; SeCSe, 0.245, 0, 1; OCS, 0.148, 0.710, 0.5; OCSe, 0.154, 0.59, 0.5; SCSe, 0.239, 0.031, 0.5; SCTe, 0.300, 0.172, 0.5.
Sets TB3b, TM3b. M^1ZM^2, α, μ; FCN, 0.052, 2.17; ClCN, 0.103, 2.802; BrCN, 0.132, 2.94; ICN, 0.182, 3.64; MeCN, 0.099, 3.97; MeNC, 0.099; 3.83.
Sets TB4a, TM4a. BX_3. X, α, μ; F, 0.034, 0; Cl, 0.187, 0; Br, 0.274, 0; I, 0.424, 0; Me, 0.175, 0.
Sets TB4b, TM4b. $X^1X^2CZ^a$. X^1, X^2, α, μ; F, F, 0.054, 0.951; Cl, Cl, 0.156, 1.17; Br, Br, 0.214, 1.17^B; H, H, 0.069, 2.339; F, H, 0.058, 2.02; Me, Me, 0.148, 2.93; Me, H, 0.105, 2.69; Cl, Cl, 0.241, 0.28^a.
Sets TB5, TM5, TC3. $MX^1X^2X^3X^4$, α, μ, ω; SiH_4, 0.111, 0, 1; SiF_4, 0.097, 0, 1; $SiCl_4$, 0.301, 0, 1; $SiBr_4$, 0.417, 0, 1; SiI_4, 0.587, 0, 1; $SiMe_4$, 0.285, 0, 1; GeH_4, 0.126, 0, 1; $GeCl_4$, 0.316, 0, 1; $GeBr_4$, 0.432, 0, 1; GeI_4, 0.632, 0, 1; $GeMe_4$, 0.300, 0, 1; SnH_4, 0.179, 0, 1; $SnCl_4$, 0.369, 0, 1; $SnBr_4$, 0.485, 0, 1; SnI_4, 0.685, 0, 1; $SnMe_4$, 0.353, 0, 1; $PbCl_4$, 0.386, 0, 1; $PbMe_4$, 0.370, 0, 1; SiH_3Br; 0.180, 1.32, 0.75; SiH_3Cl, 0.151, 1.292, 0.75; $SiBr_3H$, 0.348, 0.79, 0.25; SiH_2Cl_2, 0.191, 1.181, 0.50; $SiCl_3H$, 0.261, 0.855, 0.25; SiF_3Cl 0.148, 0.49, 0.75; SiH_3I, 0.230, 1.62, 0.75; SiF_3H, 0.077, 1.26, 0.75; $SiCl_3Me$, 0.296, 1.91, 1; $SiMe_2H_2$, 0.183, 0.75, 0.50; SiH_3Me, 0.147, 0.73, 0.75; GeH_3Br, 0.205, 1.31, 0.75; GeH_3Cl, 0.176, 2.124, 0.75; GeH_2Cl_2, 0.216, 2.22, 0.50.
Set TB6. MH_n, σ_1, α. CH_3, −0.01, 0.046; SiH_3, 0.01, 0.101; GeH_3, 0.01, 0.116; SnH_3, 0, 0.169; *NH_2*, 0.17, 0.044; PH_2, 0.18, 0.112; AsH_2, 0.14, 0.146; SbH_2, 0.15, 0.176; BiH_2, 0.15, 0.282; *OH*, 0.35, 0.018; SH, 0.28, 0.082; SeH, 0.30, 0.124; TeH, 0.26, 0.175; *F*, 0.54, −0.001; Cl, 0.47, 0.050; Br, 0.47, 0.079; I, 0.40, 0.129.
Set TB7. MMe_n. M, σ_i, α, n: C, −0.01, 0.186, 3; Si, −0.11, 0.239, 3; Ge, −0.08, 0.254, 3; Sn, −0.09, 0.307, 3; Pb, −0.10, 0.324, 3; N, 0.17, 0.145, 2; P, 0.10, 0.202, 2; As, 0.10, 0.236, 2; Sb, 0.09, 0.250, 2; Bi, 0.09, 0.353, 2; O, 0.30, 0.068, 1; S, 0.30, 0.128, 1; Se, 0.28, 0.160, 1; Te, 0.25, 0.211, 1; F, 0.54, −0.001, 0; Cl, 0.47, 0.050, 0; Br, 0.47, 0.079, 0; I, 0.40, 0.129, 0.

Note: Parameter values were taken from refs. 8, 15, 40, 46–50. Strongly hydrogen bonding groups in set TB6 are italicized.

in these compounds. There is no significant difference in the values of A or in the intercepts of Eqs. 24 and 26. Inclusion of hydrogen halides in the data set (set TB2, Table 4) gave on correlation with Eq. 25 the regression equation:

$$T_{b,M^1M^2} = 1582(\pm 85.7)\alpha + 33.8(\pm 13.4)\mu + 54.7(\pm 10.5) \qquad (27)$$

$100R^2$, 96.74; A$100R^2$, 96.51; F, 193.0; S_{est}, 25.1; S^o, 0.200; N_{dp}, 16; $r_{\alpha\mu}$, 0.220; C_α, 90.3; C_μ, 9.66.

Again, polarizability is the major factor in determining boiling point in these compounds and there is no significant difference in the values of A or in the intercepts of Eqs. 24, 26 and 27. There is also no significant difference in the values of M between Eqs. 26 and 27.

3. *Linear Triatomic Molecules*

We now consider two types of triatomic linear molecules, the CO_2 analogs $M^1{=}C{=}M^2$ (set TB3a) and the cyano and isocyano derivatives (set TB3b). In set TB3b both MeCN and MeNC are included on the assumption that the methyl group can be treated as if it were a single atom. In both of these linear sets significant dipole moments occur, thus the correlation equation used is Eq. 25. The parameter values used are in Table 5. The regression equation obtained for set TB3a is:

$$T_{b,M^1CM^2} = 1473(\pm 52.2)\alpha - 3.62(\pm 11.3) \qquad (28)$$

$100r^2$, 99.63; F, 797.8; S_{est}, 5.14; S^o, 0.0790; N_{dp}, 5; $r_{\alpha\mu}$, 0.955.

μ is strongly colinear in α.

The regression equation obtained for set TB3b is:

$$T_{b,XCN/XNC} = 864(\pm 232)\alpha + 45.9(\pm 8.84)\mu + 78.8(\pm 21.6) \qquad (29)$$

$100R^2$, 97.28; A $100R^2$, 96.37; F, 35.74; S_{est}, 12.0; S^o, 0.261; N_{dp}, 5; $r_{\alpha\mu}$, 0.436; C_α, 79.0; C_μ, 21.0.

What is most surprising in a comparison of Eqs. 28 and 29 is in the values of the coefficient A and the intercept a_o. In view of these differences sets TB3a and TB3b cannot be combined into a single data set in spite of the similarity in geometry. A for set TB3a is in good agreement with that for set TB2 but the intercepts are very different.

4. *Trigonal Planar Molecules*

We have studied the boiling points of trigonal planar molecules of two types; the boranes BX_3 (set TB4a) and the carbonyl compounds $X^1X^2C{=}O$ (set TB4b). The boranes have no dipole moment and therefore were correlated with Eq. 22. The α values used are given in Table 5. The regression equation is:

$$T_{b,BX_3} = 797(\pm 24.4)\alpha + 145(\pm 6.18) \qquad (30)$$

$100r^2$, 99.72; F, 1071; S_{est}, 6.98; S^o, 0.0682; N_{dp}, 5.

The carbonyl derivatives were correlated with Eq. 25 as they had molecular dipole moments. The parameters used are reported in Table 5. The regression equation was:

$$T_{b,X^1X^2C{=}O} = 851(\pm 55.5)\alpha + 28.8(\pm 4.23)\mu + 124(\pm 12.8) \qquad (31)$$

$100R^2$, 97.92; A $100R^2$, 97.58; F, 117.8; S_{est}, 9.21; S^o, 0.182; N_{dp}, 8; $r_{\alpha\mu}$, 0.478; C_α, 85.5; C_μ, 14.5.

As there was no significant difference in the values of A and a_o of the regression equations for sets TB4a and TB4b They were combined into a single set (set TB4, Table 4) and correlated with Eq. 25. The resulting regression equation was:

$$T_{b,trigpl} = 814(\pm 26.2)\alpha + 24.0(\pm 2.55)\mu + 139(\pm 6.50) \qquad (32)$$

$100R^2$, 99.01; A$100R^2$, 98.92; F, 498.8; S_{est}, 8.72; S^o, 0.113; N_{dp}, 13; $r_{\alpha\mu}$, 0.469; C_α, 87.2; C_μ, 12.8.

5. *Tetrahedral Molecules*

The T_b values for MX_4 compounds[36] where M is Si, Ge, Sn, or Pb (set TB5a, Table 4) were correlated with Eq. 22 using the parameters in Table 5. The regression equation is:

$$T_b = 836(\pm 20.0)\alpha + 78.0(\pm 7.87) \qquad (33)$$

$100r^2$, 99.15; F, 1251; S_{est}, 14.0; S^o, 0.0981; N_{dp}, 17.

Again, in order to include compounds in which not all the X groups are the same it is necessary to introduce a term $M\mu$, Correlation of the combined data sets (sets CP5a and CP5b) with Eq. 25 gives the regression equation:

$$T_b = 831(\pm 21.5)\alpha + 30.9(\pm 4.50)\mu + 78.8(\pm 8.21) \qquad (34)$$

$100R^2$, 98.32; A$100R^2$, 98.26; F, 788.8; S_{est}, 16.4; S^o, 0.137; N_{dp}, 30; $r_{\alpha\mu}$, 0.401; C_α, 84.3; C_μ, 15.7.

Polarizability is again the major factor in determining boiling point in these compounds. There is no significant difference in the values of A or in the intercepts of Eqs. 33 and 34.

6. Hydrides and Methyl Derivatives of Groups 14 Through 17

T_b, of the hydrides of groups 14 to 17, MH_n; excluding NH_3, H_2O, and HF; (set TB6a) were correlated with the equation:

$$T_{bX} = L\sigma_{lX} + A\alpha_X + B^o \quad (35)$$

Where $X = MH_{n-1}$, σ_l is the localized electrical effect parameter, α is a polarizability parameter. In this and the following data set it was thought more convenient to represent the dipole moment by the σ_l parameter. The regression equation is

$$T_{bX} = 142(\pm 17.8)\sigma_{lX} + 880(\pm 76.3)\alpha_X + 76.9(\pm 10.7) \quad (36)$$

$100R^2$, 94.15; F, 80.56; S_{est}, 11.0; s^o, 0.276; n, 13.

The regression equation obtained for the correlation of T_b for the methyl derivatives MMe_n, set TB7, is:

$$T_{bX} = 186(\pm 38.0)\sigma_{lX} + 818(\pm 91.1)\alpha_X + 126(\pm 22.5) \quad (37)$$

$100R^2$, 90.32; F, 60.65; S_{est}, 14.5; S^o, 0.345; n, 16.

The T_b of the hydrides including HF and H_2O (set TB6b), and including HF, H_2O, and NH_3 (set TB6c) were correlated with the equation:

$$T_b = L\sigma_l + A\alpha + H\varepsilon + a_o \quad (38)$$

where ε is a hydrogen bonding parameter equal to $-m\Delta H(g)$ for the hydrogen bonds. $\Delta H(g)$ is the enthalpy of hydrogen bond formation and m is a statistical factor that accounts for the number of hydrogen bonds formed by each molecule. The parameters required for these correlations are set forth in Table 5.

Regression equations for these sets are:

$$T_b = 135(\pm 18.0)\sigma_l + 901(\pm 78.8)\alpha + 22.6(\pm 1.38)\varepsilon + 75.6(\pm 11.2) \quad (39)$$

$100R^2$, 97.15; F, 124.8; S_{est}, 11.6; S^o, 0.197; n, 15;

and:

$$T_b = 134(\pm 16.7)\sigma_l + 899(\pm 74.4)\alpha + 22.6(\pm 1.3)\varepsilon + 76.1(\pm 10.4) \quad (40)$$

$100R^2$, 97.15; F, 136.4; S_{est}, 11.1; S^o, 0.195; n, 16.

The coefficients of Eqs. 39 and 40 are in excellent agreement with each other and do not differ significantly from those of Eq. 37.

7. *Butadienes*

As a further example consider the boiling points of 1-substituted-1,3-butadienes, 1-substituted-4-methyl-1,3-butadienes, and 2-substituted-1,3-butadienes.[5] As the only interactions expected to be significant in these data sets are dipole-dipole, dipole-induced dipole, and induced dipole-induced dipole forces, we have again dropped from the IMF equation all terms other than μ_{PhX} which represents the X-C(sp^2) bond moment and the polarizability parameter α. Correlation of the boiling points of 2-substituted-1,3-butadienes (set TB8, Table 4) with Eq. 25 gave as the best regression equation:

$$T_{bX} = 14.0(\pm 5.92)\mu_X + 703(\pm 81.4)\alpha_X + 269.27(\pm 7.20) \qquad (41)$$

$100R^2$, 96.90; A$100R^2$, 96.27; F, 62.42; S_{est}, 8.97; S_o, 0.233; n, 7.

The boiling points of the 1-substituted- and 1-substituted-4-methyl-1,3-butadienes were combined into a single data set by introducing the variable n_{Me} which takes the value 1 when there is a 4-methyl group and 0 when there is not (set TB9, Table 4), thus the correlation equation is:

$$T_{bX} = M\mu_X + A\alpha_X + B_{Me}n_{Me} + h \qquad (42)$$

The best regression obtained is:

$$T_{bX} = 24.9(\pm 6.04)\mu_X + 715(\pm 67.3)\alpha_X + 40.7(\pm 5.50)n_{Me} + 274.38(\pm 5.88) \qquad (43)$$

$100R^2$, 96.97; A$100R^2$, 96.50; F, 128.1; S_{est}, 10.6; S^o, 0.201; n, 16.

Also studied were the boiling points of the substituted 1,2-propadienes (set TB10, Table 4). The correlation equation was Eq. 25, the regression equation[37] was:

$$T_b = 38.4(\pm 6.04)\mu + 485(\pm 67.3)\alpha + 249(\pm 9.32) \qquad (44)$$

$100R^2$, 95.40; A$100R^2$, 94.75; F, 62.29; S_{est}, 12.8; S^o, 0.263; n, 9.

The parameters required for these correlations are given in Table 8. The coefficients of α and μ and the intercept are in good agreement for the 1,3-butadienes (Eqs. 41 and 43).

8. *Cohesive Energies*

A property related to boiling point that is of interest is cohesive energy. The cohesive energies of MeX at 298.15° K were taken from the compilation of Majer and Svoboda[38] (set CE1, Table 4). The data set contains no compounds capable of hydrogen bonding, therefore the IMF equation has the form:

$$(\Delta U_V^o)_X = M\mu_{MeX} + A\alpha_X + S_1\upsilon_{1X} + S_2\upsilon_{2X} + h \qquad (45)$$

as only dd, di, and ii interactions should occur. Correlation of the data set with Eq. 44 gave the regression equation:

$$(\Delta U_V^o)_X = 5.60(\pm 0.363)\mu_{MeX} + 131(\pm 6.73)\alpha_X - 7.40(\pm 1.87)\upsilon_{lX} - 7.40(\pm 3.87)\upsilon_{2X} + 8.25(\pm 2.19) \qquad (46)$$

$100R^2$, 95.44; $A100R^2$, 94.97; F, 114.7; S_{est}, 2.30; S^o, 0.232; N_{dp}, 32; $r_{\mu\alpha}$, 0.218; $r_{\mu\upsilon 1}$, 0.185; $r_{\mu\upsilon 2}$, 0.110; $r_{\alpha\upsilon 1}$, 0.143; $r_{\alpha\upsilon 2}$, 0.461; $r_{\upsilon 1\upsilon 2}$, 0.124.

If the gas phase value of the dipole moment for $MeNO_2$, 3.56, is replaced by the liquid phase value, 4.39, the results are much improved. The regression equation is:

$$(\Delta U_V^o)_X = 5.45(\pm 0.319)\mu_{MeX} + 132(\pm 6.15)\alpha_X - 7.10(\pm 1.71)\upsilon_{lX} - 6.23(\pm 3.52)\upsilon_{2X} + 7.41(\pm 2.01) \qquad (47)$$

$100R^2$, 96.22; $A100R^2$, 95.82; F, 177.9; S_{est}, 2.09; S^o, 0.211; N_{dp}, 32; $r_{\mu\alpha}$, 0.237; $r_{\mu\upsilon 1}$, 0.197; $r_{\mu\upsilon 2}$, 0.136; $r_{\alpha\upsilon 1}$, 0.143; $r_{\alpha\upsilon 2}$, 0.461; $r_{\upsilon 1\upsilon 2}$, 0.124.

There is no significant difference between the coefficients of Eqs. 46 and 47.

In all of the above data sets polarizability is the dominant factor. Dipole moments and in the case of cohesive energies steric effects make a much smaller contribution.

An examination of the A values obtained for boiling point correlations shows that they fall into two groups. The monatomic, diatomic and the triatomic CO_2 analogs (sets TB1, TB2, and TB3a respectively) have an average A value of 1425 $\pm$ 186. Although in set 3a α and μ are colinear the magnitude of M is very small in comparison with that of A in all of the sets for which the comparison can be made, thus the effect on the value of A will be quite small. By contrast, all of the other data sets studied have an average value of A of 807 $\pm$ 78.2. If we neglect the cyano and isocyano derivatives (set TB3b), a very small data set, we could argue that atoms of a molecule that are not peripheral exert a smaller polarizability effect. Intermolecular forces dependent on polarizability are very strongly dependent on the distance between interacting molecules. It should be noted that in some sets the correlations have been carried out with a values that represent the polarizability of the entire molecule whereas in other sets in which the molecule has the form XG where G is a constant skeletal group only the polarizability of the substituent is accounted for by the α_X parameter. Nevertheless, both types of data sets will give the same A value. As polarizabilities are additive to a very good approximation,

$$\alpha_{XG} = \alpha_X + \alpha_G \tag{48}$$

where α_{XG} is the polarizability of the entire molecule, α_X is that of the substituent, and α_G that of the skeletal group. Then from Eq. 24:

$$Q = A\alpha_{XG} + M\mu + a_o \tag{49a}$$

$$= A\alpha_X + A\alpha_G + M\mu + a_o \tag{49b}$$

As $A\alpha_G$ is constant.

$$Q_{XG} = A\alpha_X + M\mu + a_o^l \tag{50a}$$

$$a_o^l = a_o + A\alpha_G \tag{50b}$$

C. Melting Points

1. *Introduction*

The two basic assumptions made in the application of the IMF model to the structural dependence of melting points, T_m, are that they depend on the difference in imf between the solid and liquid states; and that they are also dependent on the way in which the molecule packs in the crystal lattice.

As the melting point represents the temperature at which solid and liquid phases are in equilibrium it follows that:

$$\Delta G_m = \Delta H_m - T_m \Delta S_m \tag{51}$$

As ΔG_m must equal zero we have:

$$T_m = \frac{\Delta H_m}{\Delta S_m} \tag{52}$$

There are two possibilities:

1. ΔS_m is variable and proportional to Δh_m.
2. ΔS_m is approximately constant, in which case

$$T_m = a\Delta H_m \tag{53}$$

Evidence has been reported which shows that the first case actually occurs.[39]

2. Monatomic Species

Again, we begin with the simplest case, that of the melting points of the rare gases (set TM1, Table 6). Using the α values reported in Table 5 correlation with Eq. 22 gave the regression equation:

$$T_m = 1106(\pm 71.5)\alpha + 45.2(\pm 6.12) \quad (54)$$

$100r^2$, 98.76; F, 239.4; S_{est}, 8.82; S^o, 0.144; N_{dp}, 5.

3. Linear Diatomic Molecules

The correlation of melting points for homonuclear diatomic molecules (set TM2a, Table 6) with Eq. 22 gave the regression equation:

$$T_{m,M} = 1413(\pm 72.8)\alpha + 28.6(\pm 7.87) \quad (55)$$

$100r^2$, 98.69; F, 377.3; S_{est}, 17.3; S^o, 0.135; N_{dp}, 7.

Table 6. Melting points used in correlations, sets TM1 to TM5

Set TM1. Rare gases. M, T_m: Ne, 24.55; Ar, 83.77; Kr, 115.95; Xe, 161.35; Rn, 202.15.
Set TM2. Diatomic molecules. M^1M^2, T_m: H_2, 14.01; F_2, 53.53; Cl_2, 172.17; Br_2, 265.95; I_2, 386.65; N_2, 63.29; O_2, 54.75; CO, 74.15; NO, 109.55; ICl, 300.35(α), 287.07(β); IBr, 315.15; ClF, 119.15; BrCl, *207.15*; BrF, 240.15d; HCl, 158.35; HBr, 184.65; HI, 222.35.
Set TM3a. Linear triatomic M^1CM^2, T_m: SCS, 161.55; SeCSe, 230.15; OCS, 134.34; SCSe, 188.15; SCTe, 219.15; OCSe, 148.75.
Set TM3b. Linear triatomic M^1ZM^2, T_m: FCN, 191.15; ClCN, 266.25; BrCN, 324.55; ICN, 419.15; MeCN, 229.35; MeNC, 228.15.
Set TM4a. Trigonal planar M_3B. M, T_m: F, 144.45; Cl, 165.85; Br, 277.15; I, 323.05; Me, 293.15.
Set TM4b. Trigonal planar $M^1M^2C{=}O$. M^1, M^2, T_m: F, F, 159.15; Cl, Cl, 169.15; F, H, 131.15; Me, Me, 177.8, Me, H, 152.15; H, H, 181.15[b].
Set TM5. $MX^1X^2X^3X^4$. $MX^1X^2X^3X^4$, T_m: SiH_4, 88.5; $SiCl_4$, 203; $SiBr_4$, 279; SiI_4, 394; $SiMe_4$, 174; GeH_4, 107; $GeCl_4$, 224; $GeBr_4$, 299; GeI_4, 407; $GeMe_4$, 185; SnH_4, 123; $SnCl_4$, 239; $SnBr_4$, 303; SnI_4, 418; $SnMe_4$, 218; $PbCl_4$, 258; $PbMe_4$, 246; SiH_3Br, 179; SiH_3Cl, 155; SiH_3I, 216; SiH_3Me, 117; $SiCl_3Me$, 195; $SiCl_3H$, 147; $SiBr_3H$, 200; SiF_3H, 142; SiF_3Cl, 135; SiH_2Cl_2, 151; $SiMe_2H_2$, 123; GeH_3Br, 241; GeH_3Cl, 221; GeH_2Cl_2, 205.

Note: Melting points were taken from refs. 32–35 unless otherwise noted. d, decomposition. a. Z=O unless otherwise noted. b. Z=S.

Inclusion of heteronuclear diatomic molecules other than hydrogen halides required a term in μ, and possibly a term in ω, the probability that the molecule approaches the crystal surface in the proper direction for binding. ω takes the values 1 for a homonuclear molecule and 0.5 for a heteronuclear molecule Thus, the correlation equation becomes:

$$T_m = A\alpha + M\mu + P\omega + a_o \tag{56}$$

The regression equation obtained is:

$$T_{m,M^1M^2} = 1382(\pm 58.9)\alpha + 36.8(\pm 7.30) \tag{57}$$

$100r^2$, 98.04; F, 549.7; S_{est}, 17.6; S^{o}, 0.152; N_{dp}, 13; $r_{\alpha\mu}$, 0.268; $r_{\alpha\omega}$, 0.132; $r_{\mu\omega}$, 0.756.

Surprisingly, there is no dependence on either μ or ω. There is no significant difference between the coefficients of Eqs. 55 and 57. Inclusion of the hydrogen halides other than HF in the data set (set TM2) gave on correlation with Eq. 56 the regression equation:

$$T_{m,M^1M^2} = 1333(\pm 62.0)\alpha + 19.9(\pm 11.1)\mu + 37.5(\pm 7.60) \tag{58}$$

$100R^2$, 97.59; A$100R^2$, 97.42; F, 263.2; S_{est}, 18.1; S^{o}, 0.172; N_{dp}, 16; $r_{\alpha\mu}$, 0.273; $r_{\alpha\omega}$, 0.129; $r_{\mu\omega}$, 0.774; C_α, 93.1; C_μ, 6.94.

There is now a borderline dependence on μ but none on ω. It must be noted however that μ and ω are significantly colinear. The values of A and a_o show no significant change.

4. *Linear Triatomic Molecules*

The linear triatomic carbon dioxide analogs, M^1CM^2 (set TM3a), usually have dipole moments and were therefore correlated with Eq. 25. The resulting regression equation is:

$$T_{m,M^1CM^2} = 654(\pm 101)\alpha + 37.6(\pm 23.4) \tag{59}$$

$100r^2$, 91.37; F, 42.37; S_{est}, 12.7; S^{o}, 0.360; N_{dp}, 5; $r_{\alpha\mu}$, 0.834.

As was the case with the boiling points of these compounds α and μ are strongly colinear.

The linear triatomic cyano/isocyano compounds (set TM3b) on correlation with Eq. 25 gave the regression equation:

$$T_{m,X(CN/NC)} = 2119(\pm 86.0)\alpha - 32.0(\pm 5.28)\mu + 144(\pm 15.3) \tag{60}$$

$100R^2$, 99.54; A $100R^2$, 99.43; F, 325.4; S_{est}, 7.28; S^{o}, 0.0958; N_{dp}, 6; $r_{\alpha\mu}$, 0.481; C_α, 93.0; C_μ, 7.02.

5. Trigonal Planar Molecules

The trigonal planar boranes (set TM4a) when correlated with Eq. 22 gave on the exclusion of the melting point for BMe_3 the regression equation:

$$T_{m,BX_3} = 469(\pm 102)\alpha + 107(\pm 27.5) \tag{61}$$

$100r^2$, 91.36; F, 21.15; S_{est}, 28.8; S^o, 0.416; N_{dp}, 4.

The fit is poor but adequate.

The trigonal planar carbonyl derivatives (set TM4b) show remarkably little variation in melting point, the range is only 29°. It is hardly surprising that when they were correlated with Eq. 25 they did not give a significant regression equation.

6. Tetrahedral Molecules

The melting points of the tetrahedral compounds $MX^1X^2X^3X^4$ where M is a group 14 element other than carbon have been modeled.[37] The melting points of the subset MX_4 where X is H, Cl, Br, I, or Me (set TM5a, Table 5) is a simple case as the dipole moment is zero for these compounds and hydrogen bonding is not possible. As all of these compounds have similar molecular geometry they are likely to have similar crystal lattices, this packing effects should be constant. Then only induced dipole–induced dipole interactions are possible and the correlation equation is Eq. 22.

The regression equation obtained is:

$$T_m = 613(\pm 19.9)\alpha + 17.3(\pm 8.05) \tag{62}$$

$100r^2$, 98.44; F, 946.5; S_{est}, 12.9; S^o, 0.133; N_{dp}, 17.

In order to include the compounds for which all four X groups are not the same it is necessary to introduce a term in the dipole moment μ. Another term that must be considered is due to the fact that while all four surfaces of the tetrahedral molecule MX_4 are equivalent those of related molecules are not. Thus, $MX^1{}_2X^2{}_2$ has two sets of two equivalent faces while $MX^1{}_3X^2$ has a set of three equivalent faces and a set of one face. We assume that the face with the largest value of $\Sigma\alpha_X$ will preferentially bind to the crystal surface. Then we assign values of the probability ω as 1 for all MX_4, 0.75 to $MX^1{}_3X^2$ when the set of three equivalent faces has the higher value of $\Sigma\alpha_X$, 0.25 when the reverse is the case, and 0.50 to $MX^1{}_2X^2{}_2$. Thus, the correlation equation becomes:

$$T_m = A\alpha + M\mu + P\omega + q_o \tag{63}$$

Correlation of the entire data set (set TM5) with Eq. 63 gave on the exclusion of the values for $SiCl_3Me$ and GeH_3Br the regression equation:

$$T_m = 594(\pm 18.7)\alpha + 52.3(\pm 5.63)\mu + 97.6(\pm 18.1)\omega - 72.2(\pm 18.5) \quad (64)$$

$100R^2$, 97.97; A $100R^2$, 97.80; F, 385.4; S_{est}, 13.5; S_o, 0.154; N_{dp}, 28; $r_{\alpha\mu}$, 0.504; $r_{\alpha\omega}$, 0.402; $r_{\mu\omega}$, 0.726; C_α, 44.2; C_μ, 19.5; C_ω, 36.3.

As a hypothetical reference compound for calculating C we have chosen that for which $\alpha = 0.2, \mu = 1$, and $\omega = 1$. Though α and μ are highly significant the 0th order partial correlation coefficient of μ with ω indicates colinearity and casts some doubt on the certainty of the dependence on ω. That dependence is likely however. Polarizability makes the greatest contribution and dipole moment the least. The coefficients of α in Eqs. 61 and 63 are not significantly different but the intercepts certainly are.

7. XG Compounds

Let us consider substrates of the type XG where either X or G or both may vary in structure. Both X and G will contribute to the intermolecular forces in the liquid and solid phases. With regard to the effects of packing there are three possible cases:

1. The packing is determined largely or entirely by G with X making a negligible contribution.
2. The packing is determined largely or entirely by X with G making a negligible contribution.
3. Both X and G make major contributions to the packing.

It has been shown that if only compounds for which G is alkyl (Ak) are studied we can derive from the IMF equation linear structure-enthalpy relationships of the form:

$$T_{m,XAK} = a_l T_{m,XAK^o} + a_o \quad (65)$$

In applying Eq. 64 the reference alkyl group, Ak° and the alkyl group of interest, Ak are held constant and X is varied. The packing energy was held approximately constant by choosing long unbranched alkyl groups for both Ak and Ak°. Sets with G equal to Ar were also successfully correlated with Eq. 9. Sets with constant X and G equal to unbranched alkyl groups with two or more C atoms were successfully correlated with an equation of the form:

$$T_m = B_C n_C + B_{ev} n_{ev} + B^o \quad (66)$$

where n_C is the number of C atoms in the Ak group and n_{ev} is a parameter which accounts for any difference in packing between unbranched Ak groups with odd or even numbers of C atoms, it takes the value 1 when the number is even and 0 when it is odd. Data sets in which X was held constant and Ak allowed to vary in both length and branching were successfully correlated with an equation of the form:

$$T_m = B_C n_C + B_{\text{syn}} n_{\text{sym}} + B_1 n_1 + B_2 n_2 + B_3 n_3 \tag{67}$$

where n_{sym} is a parameter which accounts for the symmetry of the Ak group.

8. X-C(sp^3) Compounds

In the work described above the effect on melting point of variation in X was not studied. We now consider three types of data set involving variable X:

1. G is an even numbered unbranched alkyl group with n_C not less than 4 (set TM6a).
2. G is an odd numbered unbranched alkyl group with n_C not less than 5 (set TM6b).
3. G is a branched or unbranched alkyl group with n_C not more than 5 for the former or 3 for the latter (set TM6c).

The data sets studied are presented in Table 5. They were correlated with equations derived from the imf equation. Parameter values used are reported in Tables 1, 2, and 3.

We have defined a hypothetical reference group for which

$$\sigma_l = \sigma_d = \mu = n_H = n_n = \upsilon_1, = \upsilon_2 = \upsilon_3 = n_{ev} = 1, \sigma_e = 0.1, \alpha = 0.2,$$
$$n_C = 10, p_{Ak} = 1$$

a. Unbranched Skeletal Alkyl Groups

The correlation equation used for sets TM6a and TM6b is:

$$T_{mXAk} = L\sigma_{lX} + D\sigma_{dX} + R\sigma_{eX} + M\mu_X + A\alpha_X + H_1 n_{HX} + H_2 n_{nX} + S_1\upsilon_{1X} + S_2\upsilon_{2X} + S_3\upsilon_{3X} + B_C n_{CAk} + B^{\circ} \tag{68}$$

No term in i is necessary as none of the compounds studied is ionized in the solid state. The alkyl groups require no parameterization other than polarizability, represented by n_C. The steric effects of unbranched alkyl groups with three or more C atoms are constant. The electrical effects of all alkyl groups are constant.[40] The steric effect of the X group was modeled by the segmental method. In set TM6a the T_m value for $BuCONH_2$ deviated strongly while the values for BuH, BuMe, BuCH=CH_2, and DcAc deviated somewhat. Abbreviations for alkyl groups are reported in Appendix I. On exclusion of the values for these compounds the regression equation was:

$$T_{mXAk} = -77.6(\pm 18.3)\sigma_{lX} + 119(\pm 32.9)\sigma_{eX} + 12.2(\pm 2.16)\mu_X + 20.0(\pm 2.33)n_{HX} + 5.24(\pm 1.06)n_{nX} + 27.7(\pm 9.04)\upsilon_{lX} + 9.64(\pm 0.266)n_{CAk} + 136(\pm 5.37) \tag{69}$$

$100R^2$, 92.75; $A100R^2$, 92.41; F, 226.7; S_{est}, 14.5; S^o, 0.278, n, 132; $C_{\sigma 1}$, 30.9; $C_{\sigma e}$, 4.43; C_μ, 4.87; C_{nH}, 7.98; C_{nn}, 2.09; $C_{\upsilon 1}$, 11.0; C_{nC}, 38.4.

As four of the five deviant compounds in set TM6a were butyl derivatives correlation with Eq. 68 was repeated after the exclusion of those compounds for which the alkyl group was butyl (set TM6aa). The regression equation was:

$$T_{mXAk} = -70.7(\pm 11.4)\sigma_{lX} + 105(\pm 30.5)\sigma_{eX} + 13.2(\pm 1.97)\mu_X + 20.2(\pm 2.17)n_{HX} + 4.34(\pm 1.01)n_{nX} + 23.6(\pm 8.33)\upsilon_{1X} + 8.41(\pm 0.290)n_{CAk} + 153(\pm 5.25) \tag{70}$$

$100R^2$, 91.46; $A100R^2$, 90.97; F, 159.1; S_{est}, 12.5; S^o, 0.303, n, 112. $C_{\sigma 1}$, 31.2; $C_{\sigma e}$, 4.62; C_μ, 5.81; C_{nH}, 8.92; C_{nn}, 1.92; $C_{\upsilon 1}$, 10.4; C_{nC}, 37.1.

The agreement between both the coefficients and the C_i values obtained for sets TM6a and TM6aa shows that exclusion of the values for BuX did not alter the results. As the statistics obtained for set TM6a are better than those for set TM6aa there is no justification for the exclusion of the butyl derivatives.

Correlation of set TM6b with Eq. 68 also resulted in the deviation of five compounds, PeH, PeCH=CH_2, PeCH_2OH, Pe$CONH_2$, and Hp$CONH_2$. On the exclusion of these compounds the regression equation was:

$$T_{mXAk} = -93.0(\pm 19.7)\sigma_{lX} + 23.7(\pm 10.0)\sigma_{dX} + 15.7(\pm 3.12)\mu_X - 189(\pm 54.7)\alpha_X + 23.1(\pm 4.16)n_{HX} + 5.18(\pm 1.56)n_{nX} + 66.1(\pm 13.0)\upsilon_{1X} + 56.6(\pm 24.3)\upsilon_{3X} + 8.52(\pm 0.320)n_{CAk} + 142(\pm 6.55) \tag{71}$$

$100R^2$, 92.32; $A100R^2$, 91.77; F, 147.0; S_{est}, 14.2; S^o, 0.289, n, 120. $C_{\sigma 1}$, 22.9; $C_{\sigma d}$, 5.82; C_μ, 3.86; C_α, 9.30; C_{nH}, 5.68; C_{nn}, 1.28; $C_{\upsilon 1}$, 16.3; $C_{\upsilon 3}$, 13.9; C_{nC}, 21.0.

Based on our previous work and our experience with set TM6a we have assumed that pentyl derivatives belong in set TM6b.

There are many similarities between Eqs. 70 and 71 but there are also some important differences. Set TM6a shows no dependence on σ_d, α, or υ_1

while set TM6b does. Set TM6b shows no dependence on σ_e while set TM6a does. The dependence on υ_l for set TM6b is much smaller than it is for set TM6a. The C_i values demonstrate the differences between the two sets. These differences may be due to a difference in packing between even numbered and odd numbered alkyl groups.

The possibility of combining sets TM6a and TM6b into a single data set (set TM6d) was considered. For this purpose the parameter p_{ev} which is intended to account for the difference in packing between odd numbered and even numbered unbranched alkyl groups was introduced into the correlation equation. It takes the value 0 for the former and 1 for the latter. Best results were obtained on exclusion of the values for the amides with Ak equal to Bu, Pe, Hp, and No. The regression equation was:

$$\begin{aligned} T_{mXAk} &= -86.4(\pm 10.2)\sigma_{lX} + 68.1(\pm 29.7)\sigma_{eX} + 15.6(\pm 1.64)\mu_X \\ &- 107(\pm 44.3)\alpha_X + 17.8(\pm 2.09)n_{HX} + 6.28(\pm 0.914)n_{nX} \\ &+ 53.0(\pm 9.38)\upsilon_{1X} + 29.4(\pm 13.2)\upsilon_{3X} \\ &+ 9.52(\pm 0.211)n_{CAk} - 5.76(\pm 1.97)n_{ev} + 132(\pm 4.25) \end{aligned} \tag{72}$$

$100R^2$, 91.91; A$100R^2$, 91.61; F, 280.5; S_{est}, 15.4; S^o, 0.291, n, 258. $C_{\sigma l}$, 25.6; $C_{\sigma e}$, 2.02; C_μ, 4.61; C_α, 6.33; C_{nH}, 5.28; C_{nn}, 1.86 $C_{\upsilon 1}$, 15.7; $C_{\upsilon 1}$, 8.71; C_{nC}, 28.2; C_{nev}, 1.70.

The fit of the data for the combined set is similar to that obtained for sets TM6a and TM6b. It is significant that many of the compounds which deviated from sets TM6a and TM6b could be included in set TM6c. Surprisingly, no dependence on p_{ev} was observed.

The possibility of including EtX and PrX in the data set (set TM6e) was then considered. Best results were obtained on the exclusion of T_m values for $AkCONH_2$ with Ak equal to Et, Pr, Bu, Pe, Hp, and No; of EtX with X equal to Me, $CH{=}CH_2$, CH_2OH, and Bz; and of PrH. The regression equation was:

$$\begin{aligned} T_{mXAk} &= -69.3(\pm 12.1)\sigma_{lX} + 15.8(\pm 6.06)\sigma_{dX} + 57.2(\pm 30.0)\sigma_{eX} \\ &+ 15.8(\pm 1.99)\mu_X - 93.4(\pm 44.7)\alpha_X + 22.4(\pm 2.67)n_{HX} + 5.96(\pm 1.03)n_{nX} \\ &+ 53.2(\pm 9.69)\upsilon_{1X} + 40.5(\pm 13.1)\upsilon_{3X} + 9.91(\pm 0.194)n_{CAk} \\ &+ 121(\pm 4.24) \end{aligned} \tag{73}$$

$100R^2$, 92.46; A$100R^2$, 92.23; F, 358.0; S_{est}, 16.9; S^o, 0.280, n, 303. $C_{\sigma l}$, 20.2; $C_{\sigma d}$, 4.59; $C_{\sigma e}$, 1.67; C_μ, 3.74; C_α, 5.44; C_{nH}, 6.51; C_{nn}, 1.74 $C_{\upsilon 1}$, 15.5; $C_{\upsilon 3}$, 11.8; C_{nC}, 28.9.

The fit of set TM6e to the model is comparable to that of set TM6d. Though the standard error of the estimate is somewhat larger in TM6e the other statistics are marginally better. The differences between the coefficients H_1 and B_C in the two sets are barely significant, all of the other coefficients do not differ significantly. The two sets differ in that there is no dependence of set TM6e on p_{ev} and no dependence of set TM6d on σ_d. The existence of a significant difference in packing between unbranched chains longer than propyl and those which are shorter was then studied. The packing parameters n_{Et} and n_{Pr} were introduced to account for this difference. latter and 0 otherwise. They take the value 1 when Ak is Et for the former or Pr for the latter and 0 otherwise. The correlation equation then becomes:

$$T_{mXAk} = L\sigma_{lX} + D\sigma_{dX} + R\sigma_{eX} + M\mu_X + A\alpha_X + H_1 n_{HX} + H_2 n_{nX} + S_1\upsilon_{1X} + S_2\upsilon_{2X} + S_3\upsilon_{3X} + B_C n_{CAk} + p_{Et} n_{Et} + p_{Pr} n_{Pr} + B^o \quad (74)$$

Correlation of set TM6e with Eq. 74 (set TM6ee) gave best results on the exclusion of the data points for $AkCONH_2$ with Ak equal to Et, Pr, Bu, Pe, Hp, and No; EtX with X equal to Me, $CH{=}CH_2$, CH_2OH, and Bz; and $PrCO_2H$. The groups excluded are almost the same as those excluded from set TM6e. The regression equation obtained was:

$$T_{mXAk} = -71.1(\pm 11.8)\sigma_{lX} + 12.2(\pm 5.88)\sigma_{dX} + 53.3(\pm 29.6)\sigma_{eX} + 13.6(\pm 1.93)\mu_X - 80.2(\pm 43.4)\alpha_X + 21.3(\pm 2.59)n_{HX} + 5.92(\pm 0.994)n_{nX} + 54.7(\pm 9.32)\upsilon_{1X} + 37.8(\pm 12.7)\upsilon_{3X} + 9.66(\pm 0.200)n_{CAk} - 19.7(\pm 3.82)n_{Pr} + 122(\pm 4.12) \quad (75)$$

$100R^2$, 93.12; A$100R^2$, 92.88; F, 357.8; S_{est}, 16.4; S^o, 0.268, n, 303. $C_{\sigma l}$, 20.1; $C_{\sigma d}$, 3.44; $C_{\sigma e}$, 1.50; C_μ, 3.85; C_α, 4.53; C_{nH}, 6.02; C_{nn}, 1.67; $C_{\upsilon 1}$, 15.4; $C_{\upsilon 3}$, 10.7; C_{nC}, 27.3; C_{nPr}, 5.55.

The results are comparable to but slightly better than those for set TM6e. The values of the coefficients in Eq. 75 are equivalent to those in Eq. 73. There is no dependence on n_{Et} but a highly significant dependence on n_{Pr}. The possibility of including MeX in the data set (set TM6f) by introducing the packing parameter n_{Me} into the correlation equation was then attempted. Correlation of set TM6f with a variant of Eq. 74 including n_{Me} gave best results on the exclusion of the T_m values for $AkCONH_2$ with Ak equal to Me, Et, Pr, Bu, Pe, Hp, and No; EtX with X equal to Me, $CH{=}CH_2$, CH_2OH, and Bz; PrX with X equal to CO_2H and H; and MeX with X equal to Me, Et, $CH{=}CH_2$, CH_2OH, CO_2H, and NO_2. The regression equation obtained was:

$$T_{mXAk} = -63.5(\pm 11.7)\sigma_{lX} + 13.9(\pm 5.72)\sigma_{dX} + 85.8(\pm 25.6)\sigma_{eX} + 13.0(\pm 1.92)\mu_X; +23.9(\pm 2.42)n_{HX} + 4.45(\pm 0.894)n_{nX} + 42.6(\pm 7.02)\upsilon_{1X} + 24.4(\pm 9.01)\upsilon_{3X} + 9.65(\pm 0.203)n_{CAk} - 17.7(\pm 3.97)n_{Pr} + 13.8(\pm 4.38)n_{Me} + 124(\pm 4.21) \quad (76)$$

$100R^2$, 92.93; A$100R^2$, 92.70; F, 369.2; S_{est}, 16.7; S^{o}, 0.271, n, 321. $C_{\sigma l}$, 19.7; $C_{\sigma d}$, 4.30; $C_{\sigma e}$, 2.66; C_{μ}, 4.03; C_{nH}, 7.91; C_{nn}, 1.38 $C_{\upsilon 1}$, 13.2; $C_{\upsilon 3}$, 7.56; C_{nC}, 29.9; C_{nMe}, 4.27; C_{nPr}, 5.50.

There is a significant dependence on n_{Me}. The results are comparable to those for set TM6ee.

Though the coefficients of Eq. 76 are not different from those of Eqs. 73 and 75 there is no dependence on the polarizability of X observed for set TM6f.

The packing parameterization that was introduced for individual alkyl groups has the form $p_{\text{Aki}}n_{\text{Aki}}$ where n_{Aki} is 1 when the alkyl group is Ak_{i} and 0 when it is not. The value of p_{Aki} is obtained from the regression equation. In this parameterization the variation of the packing with alkyl group structure is accounted for by p_{Ak}. Then, setting the sum of the packing terms in the regression equal to $B_{\text{P}}p_{\text{Ak}}$ we have:

$$B_P p_{Ak} = \sum_{i=1}^{n} p_{Ak,i} n_{Ak,i} = n_{Ak,i} \sum_{i=1}^{n} p_{Ak,i} \quad (77)$$

As n_{Aki} is 1 it can be seen on equating coefficients that the p_{Ak} values obtained from Eq. 76 can be used as parameters that represent the packing effect of individual alkyl groups while the coefficient B_{C} should have a value of 1. The correlation equation then becomes:

$$T_{mXAk} = L\sigma_{lX} + D\sigma_{dX} + R\sigma_{eX} + M\mu_X + A\alpha_X + H_1 n_{HX} + H_2 n_{nX} + S_1\upsilon_{1X} + S_2\upsilon_{2X} + S_3\upsilon_{3X} + B_C n_{CAk} + B_P p_{Ak} + B^o \quad (78)$$

Correlation of set TM6f with Eq. 78 after excluding the same T_{m} values (set TM6ff) gave the regression equation:

$$T_{mXAk} = -62.8(\pm 11.7)\sigma_{lX} + 13.8(\pm 5.69)\sigma_{dX} + 87.8(\pm 25.5)\sigma_{eX} + 12.9(\pm 1.92)\mu_X + 23.9(\pm 2.41)n_{HX} + 4.42(\pm 0.892)n_{nX} + 42.3(\pm 6.99)\upsilon_{1X} + 24.2(\pm 8.99)\upsilon_{3X} + 9.64(\pm 0.179)n_{CAk} + 1.02(\pm 0.165)p_{Ak} + 125(\pm 4.07) \quad (79)$$

$100R^2$, 92.95; A$100R^2$, 92.74; F, 408.6; S_{est}, 16.6; S^{o}, 0.270, n, 321. $C_{\sigma l}$, 21.6;

$C_{\sigma d}$, 4.75; $C_{\sigma e}$, 3.02; C_{μ}, 4.45; C_{nH}, 8.24; C_{nn}, 1.52; $C_{\upsilon 1}$, 14.6; $C_{\upsilon 3}$, 8.32; C_{nC}, 33.2; C_{pAk}, 0.352.

The results are almost identical to those from Eq. 76 as is shown by comparing the C_i values.

b. Branched and Short Alkyl Groups

T_m values of AkX in which Ak varies and X is constant were successfully correlated with Eq. 66. Set 6c was therefore correlated with the IMF equation in the form:

$$T_{mXAk} = L\sigma_{lX} + D\sigma_{dX} + R\sigma_{eX} + M\mu_X + A\alpha_X + H_1 n_{HX} + H_2 n_{nX} + S_1\upsilon_{1X} + S_2\upsilon_{2X} + S_3\upsilon_{3X} + B_C n_{CAk} + B_1 n_{1Ak} + B_2 n_{2Ak} + B_3 n_{3Ak} + B^o \quad (80)$$

in which steric effects of the alkyl group have been parameterized by the simple branching equation. The results were poor, accounting for only about 65% of the variance of the data. The use of a symmetry parameter based on the number of planes of symmetry including the longest chain of the hydrogen suppressed alkyl group, and a shape parameter given by:

$$p_{sh} = \frac{n_C}{(n_C - n_{C'})} \quad (81)$$

where n_C is the total number of C atoms in the alkyl group and $n_{C'}$ is the number of C atoms in branches was considered. Best results resulted from introduction of the parameter p_q which represents the number of "quasiquarternary" C atoms (those atoms that are bonded to four atoms other than hydrogen) present in the compound. The compounds $EtCH_2OH$, iPrPh, tBuI, tBuMe, neoPeH, and $neoPeCH_2OH$ deviated significantly. On excluding these compounds the regression equation obtained was:

$$T_{mXAk} = 58.2(\pm 30.0)\sigma_{lX} + 69.1(\pm 11.8)\sigma_{dX} + 340(\pm 64.9)\sigma_{eX} + 14.0(\pm 4.78)\mu_X - 385(\pm 56.6)\alpha_X + 72.3(\pm 5.37)n_{HX} + 44.9(\pm 18.2)\upsilon_{1X} + 58.8(\pm 6.37)p_{qAk} + 88.6(\pm 8.63) \quad (82)$$

$100R^2$, 84.22; $A100R^2$, 83.35; F, 84.05; S_{est}, 25.8; S^o, 0.411, n, 135. $C_{\sigma l}$, 13.6; $C_{\sigma d}$, 16.1 $C_{\sigma e}$, 7.93; C_{μ}, 3.28; C_{α}, 18.0; C_{nH}, 16.9; $C_{\upsilon 1}$, 10.5; C_{pq}, 23.7.

In order to determine whether compounds with branched alkyl groups and with unbranched alkyl groups can be included in the same data set the compounds of set TM6c which have unbranched alkyl groups (those with Ak equal to Me, Et, or Pr) (set TM6cc) were correlated with the IMF equation. The regression equation was:

$$T_{mXAk} = 95.5(\pm 38.9)\sigma_{lX} + 56.9(\pm 16.2)\sigma_{dX} + 166(\pm 84.1)\sigma_{eX} + 9.62(\pm 6.93)\mu_X - 474(\pm 68.1)\alpha_X + 72.0(\pm 7.53)n_{HX} + 83.7(\pm 10.0) \quad (83)$$

$100R^2$, 79.95; A$100R^2$, 78.57; F, 56.74; S_{est}, 26.5; S°, 0.469; n, 78. $C_{\sigma l}$, 27.6; $C_{\sigma d}$, 16.5; $C_{\sigma e}$, 4.82; C_{μ}, 2.78; C_{α}, 27.5; C_{nH}, 20.8.

Comparison of the results for Eqs. 82 and 83 shows that the coefficients of most variables are in rough agreement. The difference between the sets is that set TM6c is dependent on υ_l whereas set TM6cc is not. As the branched alkyl groups in set TM6c generally exhibit a larger steric effect than the unbranched alkyl groups the lack of dependence on υ_l is reasonable.

c. *All Alkyl Groups*

In view of the mediocre results obtained for set TM6c the possibility of including XAk with branched alkyl groups in set TM6f was considered. Unbranched alkyl groups were included with iPrX and tBuX in set TM6g. This set was correlated with a variant of Eq. 78 which included the parameters n_{iPr} and n_{tBu}. Best results were obtained on the exclusion of the points excluded from set TM6f; the T_{m} values for iPrX with X equal to H, Me, CH=CH$_2$, Ph, and CONH$_2$; those for tBuX with X equal to H, SH, and CN; and EtH. The regression equation obtained was:

$$T_{mXAk} = -60.8(\pm 12.2)\sigma_{lX} + 23.2(\pm 5.91)\sigma_{dX} + 99.3(\pm 29.6)\sigma_{eX} + 12.3(\pm 1.97)\mu_X - 97.2(\pm 43.7)\alpha_X + 24.4(\pm 2.63)n_{HX} + 4.47(\pm 0.994)n_{nX} + 60.1(\pm 9.85)\upsilon_{1X} + 43.2(\pm 11.3)\upsilon_{3X} + 9.56(\pm 0.189)n_{CAk} + 1.04(\pm 0.177)p_{Ak} + 29.7(\pm 4.36)n_{tBu} + 126(\pm 4.48) \quad (84)$$

$100R^2$, 91.59; A$100R^2$, 91.32; F, 306.8; S_{est}, 17.9; S°, 0.296, n, 351. $C_{\sigma l}$, 15.9; $C_{\sigma d}$, 6.04; $C_{\sigma e}$, 2.58; C_{μ}, 3.20; C_{α}, 5.06; $C_{n\text{H}}$, 6.35; C_{nn}, 1.16 $C_{\upsilon 1}$, 15.6; $C_{\upsilon 3}$, 11.3; C_{nC}, 24.9; C_{pAk}, 0.271; C_{ntBu}, 7.73.

The results are not quite as good as those obtained for set TM6f though the coefficients are generally in good agreement. Like sets TM6e and TM6ee but unlike set TM6f there is some dependence on polarizability. No dependence on n_{iPr} was observed. Correlation of set TM6g with Eq. 78 using p_{Ak} values of 0 and 29.7 for iPr and tBu respectively (obtained from Eq. 84) on exclusion of the same T_{m} values as in set TM6g gave the regression equation (set TM6gg):

$$T_{mXAk} = -59.9(\pm 12.1)\sigma_{lX} + 23.7(\pm 5.88)\sigma_{dX} + 110(\pm 29.4)\sigma_{eX} + 12.6(\pm 1.96)\mu_X - 85.6(\pm 43.4)\alpha_X + 25.5(\pm 2.61)n_{HX} + 4.21(\pm 0.990)n_{nX} + 58.0(\pm 9.80)\upsilon_{1X} + 41.5(\pm 11.8)\upsilon_{3X} + 9.57(\pm 0.186)n_{CAk} + 1.06(\pm 0.113)p_{Ak} + 127(\pm 4.44) \quad (85)$$

$100R^2$, 91.67; A$100R^2$, 91.43; F, 339.2; S_{est}, 17.8; $S°$, 0.294, n, 351. $C_{\sigma l}$, 17.1; $C_{\sigma d}$, 6.78; $C_{\sigma e}$, 3.14; C_{μ}, 3.59; C_{α}, 4.89; C_{nH}, 7.27; C_{nn}, 1.20 $C_{\upsilon 1}$, 16.6; $C_{\upsilon 3}$, 11.9; C_{nC}, 27.3; C_{pAk}, 0.302.

The results for set TM6gg are comparable to those for set TM6g. The T_m values for iBuX and neoPeX were then included in set TM6g (set TM6h). All of the T_m values excluded from set TM6g and the values for $iBuCONH_2$ and $neoPeC_2H$ were excluded from set TM6h. The regression equation was:

$$T_{mXAk} = -64.4(\pm 12.0)\sigma_{lX} + 22.9(\pm 5.85)\sigma_{dX} + 106(\pm 29.1)\sigma_{eX} + 13.6(\pm 1.93)\mu_X - 93.9(\pm 42.9)\alpha_X + 25.4(\pm 2.60)n_{HX} + 4.51(\pm 0.977)n_{nX} + 61.9(\pm 9.63)\upsilon_{1X} + 45.1(\pm 11.6)\upsilon_{3X} + 9.57(\pm 0.187)n_{CAk} + 1.05(\pm 0.114)p_{Ak} - 22.4(\pm 5.60)n_{iBu} + 96.2(\pm 9.17)n_{neoPe} + 124(\pm 4.31) \quad (86)$$

$100R^2$, 91.85; A$100R^2$, 91.58; F, 305.3; S_{est}, 17.9; $S°$, 0.291, n, 366. $C_{\sigma l}$, 13.4; $C_{\sigma d}$, 4.75; $C_{\sigma e}$, 2.21; C_{μ}, 2.81; C_{α}, 3.89; C_{nH}, 5.26; C_{nn}, 0.935; $C_{\upsilon 1}$, 12.8; $C_{\upsilon 3}$, 9.35; C_{nC}, 19.8; C_{pAk}, 0.218; C_{niBu}, 4.63; C_{nneoPe}, 19.9.

Correlation of set TM6h with Eq. 78 after excluding the same T_m values excluded from set TM6g gave the regression equation (set TM6gg):

$$T_{mXAK} = -64.4(\pm 12.0)\sigma_{lX} + 22.9(\pm 5.82)\sigma_{dX} + 105(\pm 28.9)\sigma_{eX} + 13.6(\pm 1.92)\mu_X - 94.5(\pm 42.8)\alpha_X + 25.3(\pm 2.59)n_{HX} + 4.52(\pm 0.973)n_{nX} + 62.1(\pm 9.59)\upsilon_{1X} + 45.5(\pm 11.5)\upsilon_{3X} + 9.57(\pm 0.184)n_{CAk} + 1.02(\pm 0.0693)p_{Ak} + 124(\pm 4.24) \quad (87)$$

$100R^2$, 91.85; A$100R^2$, 91.62; F, 362.7; S_{est}, 17.9; $S°$, 0.290, n, 366. $C_{\sigma l}$, 17.7; $C_{\sigma d}$, 6.28; $C_{\sigma e}$, 2.89; C_{μ}, 3.72; C_{α}, 5.19; C_{nH}, 6.95; C_{nn}, 1.24 $C_{\upsilon 1}$, 17.0; $C_{\upsilon 3}$, 12.5; C_{nC}, 26.3; C_{pAk} 0.280. The correlation matrix for this equation is given in Table 8.

Apparently, all types of alkyl groups can be included in the correlation by determining P_{Ak} values for them.

Table 7. Melting points for set TM6

Ak: X, T_m (deg. K):
Me: F, 131.15; Cl, 175.45; Br, 189.15; I, 206.65; SH, 150.15; CO_2Et, 189.55; CO_2Me, 175.05; CN, 229.35; C_2H, 170.35; Ac, 177.80; CHO, 149.65; OAc, 175.05; OH, 179.25; NH_2, 179.65; Ph, 178.15; OMe, 134.65; SMe, 174.85; SEt, 167.25; COCl, 160.25; Me, 89.85; Et, 85.45; Vi, 87.95; NO_2, 256.15; $CONH_2$, 353.25; CO_2H, 289.78; CH_2OH, 158.65.
Et: H, 89.85; F, 129.95; Cl, 136.15; Br, 154.55; I, 162.25; SH, 125.25; CO_2Et, 199.25; CO_2Me, 185.15; CN, 180.35; C_2H, 147.45; Ac, 186.45; Me, 85.45; Et, 134.75; CHO, 192.15; OAc,189.15; CO_2H, 252.35; OH, 158.65; NH_2, 192.15; Ph,178.15; OEt, 156.85; SMe, 167.25; COCl, 179.15; NO_2, 183.15; Vi, 87.85; Bz, 291.75; $CONH_2$, 352.15; CH_2OH, 146.65.
Pr: H, 85.45; Me, 134.75; Et, 143.45; F, 114.15; Cl, 150.35; Br, 163.05; I, 172.15; SH, 160.05; CO_2Et, 175.15; CO_2Me, 188.35; CN, 161.25; C_2H, 141.25; Vi, 107.95; Ac, 195.35; CHO, 176.75; OAc, 181.15; OH, 146.65; CH_2OH, 184.55; NH_2, 190.15; Ph, 173.55; SMe, 160.15; COCl, 184.15; NO_2, 181.85; $CONH_2$, 389.15; CO_2H, 268.15.
iPr: H, 85.45; Me, 113.75; Cl, 155.95; Br, 184.15; I, 183.05; SH, 142.65; CO_2Me, 188.45; CN, 201.65; C_2H, 183.45; Vi, 104.65; Ac, 181.15; OAc, 199.75; CO_2H, 243.85; OH, 183.65; NH_2, 177.95; $CONH_2$, 401.15; Ph, 249.95; NO_2, 180.15.
Bu: H, 134.75; Me, 143.45; Et, 177.75; F, 139.15; Cl, 150.05; Br, 160.75; I, 169.65; SH, 157.45; CO_2Et, 181.85; CN, 176.35; C_2H, 167.45; Vi, 133.35; Ac, 216.25; CHO, 181.15; CO_2H, 239.45; OH, 184.55; CH_2OH, 194.25; NH_2, 190.15; Ph, 173.55; COCl, 163.15; OAc, 199.65; OMe, 157.65; OEt, 170.15; NO_2, 191.85; $CONH_2$, 387.15.
iBu: H, 113.75; Cl, 142.85; Br, 155.75; CN, 172.35; CO_2Et, 183.65; CHO, 222.15; Ac, 188.45; C_2H, 168.05; Vi, 119.55; $CONH_2$, 410.15; OAc, 174.57; SCN, 214.15.
tBu: H, 113.75; Me, 256.65; Cl, 247.75; Br, 256.95; I, 234.95; SH, 274.26; CO_2Et, 183.65; CN, 288.65; C_2H, 191.95; Vi, 157.95; Ac, 223.35; CO_2H, 308.15; OH, 247.65; NH_2, 205.65; CHO, 279.15; OMe, 164.15; OEt, 179.15; SEt, 182.15; SMe, 175.35; SCN, 283.65; Ph, 215.35.
Pe: H, 143.45; Me, 177.75; Et, 182.55; Cl, 174.15; Br, 185.15; I, 187.15; SH, 197.45; CO_2Et, 206.15; CO_2Me, 202.15; CN, 192.85; C_2H, 192.15; Vi, 154.25; Ac, 238.15; CHO, 217.15; CO_2H, 269.15; OH, 194.25; CH_2OH, 221.55; NH_2, 218.15; Ph, 194.85; OAc, 202.35; COCl, 186.15; $CONH_2$, 389.15.
neoPe: H, 256.65; Cl, 253.15; C_2H,197.45; $CONH_2$, 407.15; OH, 325.65.
Hx: H, 177.75; Me, 182.55; Et, 216.35; Br, 188.15; SH, 192.65; CO_2Et, 207.15; C_2H, 193.85; Vi, 171.45; Ac, 257.15; CHO, 230.15; CO_2H, 265.65; OH, 221.55; CH_2OH, 238.55; NH_2, 250.15; OAc, 192.25.
Hp: H, 182.55; Me, 216.35; Et, 219.65; Cl, 204.15; Br, 215.15; I, 224.95; SH, 229.95; CO_2Et, 226.15; CO_2Me, 233.15; CN, 227.55; C_2H, 223.15; Vi, 191.75; Ac, 265.65; CO_2H, 289.75; OH, 238.55; CH_2OH, 258.15; NH_2, 250.15; OAc, 222.95; COCl, 210.15; $CONH_2$, 381.15.

(continued)

Table 7. Continued

Oc: H, 216.35; Me, 219.65; Et, 243.45; Cl, 212.15; Br, 218.15; I, 227.25; SH, 223.95; CN, 238.95; Vi, 206.85; CO_2H, 285.65; OH, 258.15; CH_2OH, 267.15; NH_2, 270.15; OAc, 258.10; COCl, 212.65.
No: H, 219.65; Me, 243.45; Et, 247.55; Cl, 233.75; CO_2Et, 253.15; CO_2Me, 255.15; CN, 255.25; C_2H, 248.15; Vi, 223.95; Ac, 284.65; CHO, 268.15; CO_2H, 304.55; OH, 267.65; CH_2OH, 280.05; NH_2, 272.15; COCl, 238.65; $CONH_2$, 381.15.
Dc: H, 243.45; Me, 247.55; Et, 263.55; Pr, 267.65; F, 238.15; Cl, 241.85; Br, 243.95; I, 257.85; SH, 247.15; CO_2Et, 258.15; C_2H, 254.15; Vi, 237.92; Ac, 294.15; CHO, 269.15; CO_2H, 301.75; OH, 280.15; CH_2OH, 289.15; NH_2, 288.15; CH_2CN, 277.15; OAc, 258.10.
Udc: H, 247.55; Me, 263.55; Et, 267.75; CO_2Et, 271.35; CO_2Me, 278.35; CN, 277.15; Vi, 250.05; Ac, 303.65; CO_2H, 317.75; OH, 289.05; CH_2OH, 296.95; NH_2, 290.15; $CONH_2$, 383.15; COCl, 256.15.
Ddc: H, 263.55; Me, 267.75; Et, 279.05; Cl, 263.85; Br, 263.65; I, 273.45; CO_2Me, 279.65; CN, 282.85; C_2H, 279.65; Vi, 260.25; CO_2H, 317.65; OH, 296.95; CH_2OH, 305.65; NH_2, 301.45; $CONH_2$, 373.15.
Trd: H, 267.75; Me, 279.05; Et, 283.05; Cl, 279.35; CN, 292.35; C_2H, 283.15; Vi, 278.15; Ac, 312.65; CO_2H, 331.65; OH, 305.65; CH_2OH, 310.95; NH_2, 300.55; CO_2Et, 285.45; CO_2Me, 292.15; $CONH_2$, 379.15; COCl, 272.15; CHO, 303.15.
Ted: H, 279.05; Me, 283.05; Et, 291.35; Cl, 278.75; C_2H, 288.15; Vi, 277.25; OH, 310.95; NH_2, 314.15.
Pdc: H, 283.05; Me, 291.35; Et, 295.15; Br, 292.15; CN, 304.15; Vi, 284.35; Ac, 321.15; CO_2H, 336.65; CH_2OH, 323.15; NH_2, 310.45; CO_2Et, 297.15; CO_2Me, 303.15; $CONH_2$, 380.15; COCl, 285.15; CHO, 307.15.
Hxd: H, 291.35; Me, 295.15; Et, 301.33; F, 291.15; Cl, 291.05; Br, 291.15; I, 297.85, SH, 292.15; CO_2Et, 301.15; CO_2Me, 303.15; CN, 307.15; C_2H, 300.15; Vi, 290.65; Ac, 297.35; CHO, 309.15; CH_2Br, 305.15; CO_2H, 334.35; OH, 322.42; CH_2OH, 327.15; NH_2, 319.92; $CONH_2$, 381.35; Ph, 289.25.
Hpd: H, 295.15; Me, 301.35; Et, 305.05; Cl, 305.15; CN, 314.15; Vi, 310.65; Ac, 330.15; CO_2H, 344.35; CH_2OH, 331.05; NH_2, 322.15; CO_2Et, 305.15; CO_2Me, 312.15; $CONH_2$, 381.65; COCl, 296.15.
Ocd: H, 301.33; Me, 305.05; Et, 309.95; Cl, 301.75; Br, 301.35; I, 307.15; SH, 301.15; CN, 316.15; C_2H, 309.15; Vi, 301.65; Ac, 331.15; CO_2H, 342.55; OH, 332.75; CH_2OH, 335.65; NH_2, 326.01; OAc, 307.65; NMe_2, 296.04.
Ndc: H, 305.05; Me, 309.55; CO_2H, 350.15; CO_2Et, 323.15; CO_2Me, 327.65; OH, 335.65.

Notes: Sets studied: TM6a, even numbered unbranched alkyl groups with n_C equal to 4–18; TN6b, odd numbered unbranched alkyl groups with n_C equal to 5–19; TM6c. Me, Et, Pr, iPr, iBu, tBu, iPe, tPe, neoPe; TM6d, unbranched alkyl groups with n_C equal to 4–19; TM6e, set TM6d + Et, Pr; TM6f, unbranched alkyl groups with n_C equal to 1–19; TM6g, set TM6f + iPr, tBu; TM6h, set TM6g + iBu, neoPe.
Melting points were taken from refs. 32–35.

Table 8. Parameter values used in the correlations of sets TB7 to TB10, TM6, and TM9 through TM12

X	σ_l	σ_d	σ_e	$\mu(XCsp^3)$	$\mu(XCsp^2)$	α	n_H	n_n	n_{hb}	υ_1	υ_2	υ_3
H	0	0	0	0	0	0	0	0	0	0	0	0
Group 14 X												
Me	−0.01	−0.14	−0.030	0	0.37	0.046	0	0	0	0.52	0	0
Et	−0.01	−0.12	−0.036	0	0.37	0.093	0	0	0	0.52	0.52	0
Pr	−0.01	−0.15	−0.036	0	0.37	0.139	0	0	0	0.52	0.52	0.52
iPr	0.01	−0.15	−0.040	0	0.40	0.140	0	0	0	0.76	0.52	0
Bu	−0.01	−0.15	−0.036	0	0.37	0.186	0	0	0	0.52	0.52	0.52
iBu	−0.01	−0.14	−0.036	0	0.37	0.186	0	0	0	0.52	0.76	0.52
sBu	−0.01	−0.14	−0.036	0	0.40	0.186	0	0	0	0.76	0.52	0.52
tBu	−0.01	−0.15	−0.036	0	0.52	0.186	0	0	0	1.24	0.52	0
Pe	−0.01	−0.16	−0.040	0	0.37	0.232	0	0	0	0.52	0.52	0.52
CH_2tBu	0.00	−0.14	−0.036	0	0.37	0.232	0	0	0	0.52	1.24	0.52
cHx	0.00	−0.14	−0.036	0	0.37	0.257	0	0	0	0.70	0.52	0.52
CH_2Br	0.20	−0.08	−0.026	2.069	1.87	0.124	0	0	0	0.52	0.65	0
CH_2Cl	0.17	−0.06	−0.024	1.895	1.83	0.095	0	0	0	0.52	0.55	0
CH_2F	0.21	−0.04	−0.004	1.94	1.77	0.044	0	0	0	0.52	0.27	0
CH_2I	0.17	−0.09	−0.056	1.71	1.60	0.176	0	0	0	0.52	0.78	0
CH_2NH_2	0.06	−0.12	−0.010	1.225	1.38	0.081	2	1	1	0.52	0.35	0
CH_2CN	0.20	−0.01	−0.011	3.53	3.43	0.099	0	1	0	0.52	0.40	0.40
CH_2OMe	0.11	−0.10	−0.041	1.32	1.71	0.114	0	2	0	0.52	0.32	0.52
CH_2OH	0.11	−0.10	−0.025	1.58	1.71	0.062	1	2	1	0.52	0.32	0
CH_2CO_2H	0.16	−0.09	−0.020	1.70	1.66	0.105	1	4	2	0.52	0.50	0.32
CH_2Vi	0.02	−0.16	−0.039	0.5	0.438	0.135	0	0	0	0.52	0.57	0.57

(continued)

Table 8. Continued

X	σ_I	σ_d	σ_e	$\mu(XCsp^3)$	$\mu(XCsp^2)$	α	n_H	n_n	n_{hb}	υ_1	υ_2	υ_3
CH_2Ph	0.03	−0.13	−0.057	0.37	0.22	0.290	0	0	0	0.52	0.57	0.57
$CH_2CH_2CO_2H$	0.07	−0.13	−0.020	1.23	1.65	0.151	1	4	2	0.52	0.52	0.50
CH_2CH_2Ph	0.02	−0.13	−0.018	0.55	0.21	0.336	0	0	0	0.52	0.52	0.57
$CHPh_2$	0.08	−0.14	−0.079		0.3	0.530	0	0	0		0.57	0.57
CF_3	0.40	0.13	−0.026	2.321	2.86	0.040	0	0	0	0.90	0.27	0
CCl_3	0.36	0.10	−0.018	1.755	2.14	0.191	0	0	0	1.38	0.55	0
Vi	0.11	−0.08	−0.12	0.364	0.13	0.100	0	0	0	0.57	0.57	0
1-MeVn	0.07	−0.18	−0.089	0.503	0.77	0.135	0	0	0	0.57	0.57	0
E-2-MeVn	0.10	−0.10	−0.078	0	0.72	0.135	0	0	0	0.57	0.57	0.52
E-2-VnPh	0.13	−0.33	−0.12	0.72	0	0.331	0	0	0	0.57	0.57	0.57
E-2-VnCO_2H	0.22	−0.02	−0.11	2.13	2.04	0.149	1	4	2	0.57	0.57	0.50
Ph	0.12	−0.12	−0.12	0.37	0	0.243	0	0	0	0.57	0.57	0.57
C_2H	0.29	−0.02	−0.10	0.75	0.71	0.085	1	1	0	0.58	0.58	0
CHO	0.30	0.27	−0.10	2.69	2.92	0.059	0	2	0	0.50	0	0
Ac	0.30	0.25	−0.095	2.89	2.88	0.114	0	2	0	0.50	0.52	0
EtCO	0.30	0.25	−0.095	2.775	2.90	0.149	0	2	0	0.50	0.52	0.52
PrCO	0.30	0.25	−0.095	2.775[a]	2.90[a]	0.195	0	2	0	0.50	0.52	0.52
Bz	0.30	0.22	−0.11	2.96	2.96	0.293	0	2	0	0.50	0.57	0.57
COF	0.46	0.31	−0.070	2.96		0.055	0	0	0	0.50	0.27	0
COCl	0.44	0.31	−0.060	*2.71*	3.18	0.116	0	2	0	0.50	0.32	0
CO_2H	0.30	0.17	−0.051	1.70	1.86	0.059	1	4	2	0.50	0.32	0
CO_2Me	0.32	0.16	−0.070	1.706	1.92	0.118	0	4	0	0.50	0.32	0.52
CO_2Et	0.30	0.18	−0.064	1.84	1.849	0.164	0	4	0	0.50	0.32	0.52
CO_2Bu	0.31	0.17	−0.067	1.84[a]	1.849[a]	0.258	0	4	0	0.50	0.32	0.52
$CONH_2$	0.28	0.12	−0.055	3.72	3.42	0.088	2	3	2	0.50	0.32	0

$CONEt_2$	0.28	0.05	−0.06	3.81[a]		0.278	0	3	0	0.50	0.32	0
CN	0.57	0.12	−0.055	3.9185	4.14	0.053	0	1	0	0.40	0.40	0
Group 15 X												
NH_2	0.17	−0.68	−0.13	1.296	1.95	0.044	2	1	1	0.35	0	0
NHOH	0.15	−0.39	−0.10			0.059	2	3	2	0.35	0.32	0
NCO	0.34	−0.17	0.070	2.81	2.25	0.078	0	3	0	0.35	0.40	0.32
NCS	0.54	−0.11	−0.090	3.84	2.69	0.162	0	1	0	0.35	0.40	0.60
$NHCONH_2$	0.23	−0.45	−0.13			0.135	3	4	2	0.35	0.50	0.35
NHMe	0.13	−0.67	−0.18	1.01	1.68	0.093	1	1	1	0.35	0.52	0
NHAc	0.28	−0.35	−0.088		3.75	0.139	1	3	2	0.35	0.50	0.32
NMe_2	0.17	−0.66	−0.24	0.612	1.60	0.145	0	1	0	0.63	0.52	0
$NHCO_2Et$	0.28	−0.42	−0.13		3.69	0.206	1	4	2	0.35	0.50	0.32
NHPh	0.20	−0.62	−0.18	1.67	1.14	0.290	1	1	1	0.35	0.57	0.57
NHBz	0.28	−0.39	−0.15		3.66	0.335	1	3	2	0.35	0.50	0.57
$NHNH_2$	0.11	−0.64	−0.11	*1.75*	1.67	0.074	3	2	2	0.35	0.35	0
E—N=NPh	0.27	0.12	−0.059		0	0.303	0	2	0	0.35	0.35	0.57
N_3	0.43	−0.27	−0.12	2.17	1.56	0.092	0	2	0	0.35	0.35	0.35
NO_2	0.67	0.18	−0.077	3.56	4.28	0.063	0	4	0	0.35	0.32	0
$POCl_2$	0.60	0.46	0.087			0.197	0	2	0	1.21	0.55	0
$PO(OH)_2$	0.43	0.21	−0.011			0.133	2	6	2	1.04	0.32	0
$AsO(OH)_2$	0.43	0.20	−0.003			0.220	2	6	2	1.19	0.32	0
Group 16 X												
OH	0.35	−0.57	−0.044	1.77	1.40	0.018	1	2	1	0.32	0	0
OMe	0.30	−0.55	−0.064	1.31	1.36	0.068	0	2	0	0.32	0.52	0
OEt	0.28	−0.55	−0.070	1.22	1.38	0.114	0	2	0	0.32	0.52	0.52
OAc	0.38	−0.24	−0.005	1.706	1.69	0.114	0	4	0	0.32	0.50	0.32

(*continued*)

Table 8. Continued

X	σ_I	σ_d	σ_e	$\mu(XCsp^3)$	$\mu(XCsp^2)$	α	n_H	n_n	n_{hb}	υ_1	υ_2	υ_3
OPr	0.28	−0.55	−0.067	1.27[a]	1.37[a]	0.160	0	2	0	0.32	0.52	0.52
OiPr	0.27	−0.55	−0.067	1.27[a]	1.37[a]	0.160	0	2	0	0.32	0.76	0.52
OBu	0.28	−0.55	−0.067	1.27[a]	1.37[a]	0.206	0	2	0	0.32	0.52	0.52
OPh	0.40	−0.57	−0.083	1.36	1.13	0.267	0	2	0	0.32	0.57	0.57
OBz	0.43	−0.28	−0.025	1.92	1.88	0.307	0	4	0	0.32	0.50	0.57
SH	0.27	−0.40	−0.098	*1.52*	1.21	0.082	0	0	0	0.60	0	0
SMe	0.30	−0.38	−0.13	*1.06*	1.29	0.128	0	0	0	0.60	0.52	0
SEt	0.26	−0.39	−0.12	*1.06*	1.29[a]	0.174	0	0	0	0.60	0.52	0.52
SCN	0.56	−0.15	−0.040	*4.13*	3.62	0.124	0	0	0	0.60	0.40	0.40
SPh	0.31	−0.34	−0.17	1.50	1.37	0.333	0	0	0	0.60	0.57	0.57
SSMe	0.41	−0.45	−0.17	1.985		0.210	0	0	0	0.60	0.60	0.52
SSPh	0.41	−0.43	−0.17		1.82	0.415	0	0	0	0.60	0.60	0.57
SOPh	0.51	−0.02	−0.052		4.02	0.320	0	2	0			
SO_2Me	0.59	0.13	−0.52	4.50	4.73	0.125	0	4	0	1.03	0.32	0
SO_2Ph	0.56	−0.08	−0.082	4.73	5.00	0.322	0	4	0	1.03	0.57	0.57
SO_2Cl					4.59	0.131	0	0	0	1.03	0.55	0
SO_2OH					3.80	0.099	1	6	2	1.03	0.32	0
SO_2NH_2	0.44	0.23	−0.082	4.60	5.13	0.113	2	4	2	1.03	0.35	0
SeCl	0.49	−0.36	−0.082		0.73	0.169	0	0	0	0.70	0.55	0
SeBr	0.49	−0.35	−0.084		0.76	0.193	0	0	0	0.70	0.65	0
SePh	0.26	−0.35	−0.15	1.50	1.39	0.362	0	0	0	0.70	0.57	0.57
Se_2Ph					1.37	0.476	0	0	0	0.70	0.70	0.57

Group 17 X												
F	0.54	−0.48	0.041	1.8549	1.66	−0.001	0	0	0	0.27	0	0
Cl	0.47	−0.28	−0.011	1.895	1.70	0.050	0	0	0	0.55	0	0
Br	0.47	−0.27	−0.018	1.84	1.70	0.079	0	0	0	0.65	0	0
I	0.40	−0.20	−0.057	1.618	1.71	0.129	0	0	0	0.78	0	0

Notes: Parameter values were taken from refs. 8, 15, 40, 46–50. Estimated values of μ are given in italics.
[a] Estimated value.
n. Not included in the correlation.

d. *The Prediction of* T_m

The predictive capability of the model was tested by calculating values of T_m for a test set of 45 compounds which were not included in the correlation used to obtain Eq. 87. The calculated and observed values of T_m; the differences between calculated and observed values, Δ; and the ratio of Δ to the standard error of the estimate are presented in Table 7. The T_m values of the test set fall into three groups:

1. Those whose X and Ak groups were in the data set used to obtain Eq. 87.
2. Those with X groups not included in that data set.
3. Those with Ak groups not included in that data sets.

The unbranched compounds of group 1(TedCHO, $TedCO_2H$, $Ted-CO_2Me$, OcdPh, OcdAc) all show good agreement between calculated and observed T_m values. Of the compounds with branched alkyl groups iPrSMe, tBuF, and iBuX with X equal to I, OH, NH_2, and CO_2H show good agreement, iBuSH deviates slightly, and iBuPh strongly. The neopentyl derivatives studied all show very large deviations. Of the ten neoPeX values available four (X = H,Cl, OH, $CONH_2$) seem to require p_{Ak} equal to 96.2, the remainder (X = Me, Et, Vi, C_2H, CH_2OH, and CO_2H) seem to require p_{Ak} equal to about −14.

The five compounds in Group 2 are MeX with X equal to SSMe, $NHNH_2$, CCl_3, CF_3, and SPh. The first three compounds show good agreement between observed and calculated T_m values, the fourth shows some deviation, and the last deviates severely.

Group 3 consists of 22 compounds with unbranched alkyl groups from 20 to 30 carbon atoms in length. The results show good agreement with chain lengths up to 24 C atoms. The magnitude of the deviation then increases with increasing chain length. It will be necessary to determine p_{Ak} values for the chains with ≥ 25 C atoms if they are to be included in the model.

e. *Composition of the Structural Effects*

In interpreting the results it is necessary to recall that electrical and steric effects, bond moments, and charge transfer interactions of unbranched alkyl groups with four or more carbon atoms are either constant or zero. Only polarizability effects should occur with these groups. For the other alkyl groups any other variation in the alkyl group is accounted for by p_{Ak}. The overall effect of the alkyl group, C_{Ak} is therefore given by the sum of C_{nC}, C_{nev}, and C_{pAk} in all sets which do not have terms in n_{Ak}. The values for C_{Ak} for sets TM6d, TM6ff, TM6gg, and TM6hh are 28.9, 33.6, 27.6, and 26.6 respectively. Thus, the structural effects of the substituent X are

Table 9. Correlation matrix for set TM6

σ_l	σ_d	σ_e	μ	α	n_H	n_n	υ_1	υ_2	υ_3	n_C	p_{Ak}	
1	0.071	0.106	0.788	0.056	0.127	0.255	0.198	0.082	0.026	0.187	0.009	σ_l
	1	0.006	0.321	0.310	0.381	0.411	0.119	0.387	0.129	0.073	0.076	σ_d
		1	0.021	0.397	0.223	0.040	0.308	0.316	0.034	0.050	0.003	σ_e
			1	0.053	0.073	0.414	0.102	0.019	0.078	0.081	0.003	μ
				1	0.318	0.204	0.558	0.552	0.493	0.108	0.021	α
					1	0.232	0.178	0.181	0.144	0.130	0.071	n_H
						1	0.118	0.243	0.015	0.005	0.018	n_n
							1	0.183	0.029	0.198	0.038	υ_1
								1	0.343	0.027	0.040	υ_2
									1	0.170	0.050	υ_3
										1	0.087	n_C
											1	p_{Ak}

predominant in determining the value of the melting point. The effect of the alkyl group is due largely to its polarizability.

Due to colinearities between variables it is useful to group the parameters of the IMF model into two categories:

1. Polar parameters; including σ_1, σ_d, σ_e, μ, n_H, n_n, i.
2. Nonpolar parameters; including α, υ_1, υ_2, υ_3.

The polar and nonpolar contributions, C_P and C_{NP} respectively, are given by the equations:

$$C_P = C_{\sigma l} + C_{\sigma d} + C_{\sigma e} + C_\mu + C_{nH} + C_{nn} + C_i \tag{88}$$

$$C_{NP} = C_\alpha + C_{\upsilon 1} + C_{\upsilon 2} + C_{\upsilon 3} \tag{89}$$

For sets TM6d, TM6ff, TM6gg, and TM6hh the values of C_P and C_{NP} are 38.5, 32.7; 43.6, 22.9; 39.1, 32.8; and 38.8, 34.7. The contributions are roughly comparable.

f. Packing in Monosubstituted Alkanes

In the crystal the molecule presumably adopts the conformation which maximizes intermolecular attractions. In the liquid state the molecule may have other conformations, and in any event will have a different degree of intermolecular attraction. There are therefore two different contributions to the energy change resulting from the change in state:

1. The difference in imf between the liquid and the solid states.
2. The difference in energy between the conformations in the liquid and solid states. This is the packing energy.

One of the major assumptions in this work is that the entropy is either constant or proportional to the enthalpy. This must be the case in order to fit the data to the IMF model. The packing contribution to the melting point must be dependent in some manner on ΔS. It seems reasonable then that the packing parameter p_{Ak} should be a function of ΔS. Thus we may separate ΔG_m into two components, ΔG_{imf} which accounts for the effect of intermolecular forces and ΔG_{pk} which accounts for the effect of packing:

$$\Delta G_m = \Delta G_{imf} + \Delta G_{pk} \tag{90}$$

If ΔS_{pk} depends only on packing and is constant for a given type of packing while ΔS_{imf} is constant then we may rewrite Eq. 52 as:

$$T_m = \frac{\Delta H_{\text{imf}}}{\Delta S_{\text{imf}}} + \frac{\Delta H_{pk}}{\Delta S_{pk}} \quad (91)$$

$$= f(\text{imf}) + \text{B}_{\text{Ak}}\text{p}_{\text{Ak}} \quad (92)$$

It is useful to separate packing effects into those due to the alkyl group and those due to the substituent.

g. *Alkyl Group Effects*

As the length of the alkyl group increases both the magnitude of the dispersion force and the number of possible conformations increases. These two quantities should have opposite effects on the melting point. It seems not unreasonable then that a plot of melting point versus p_{Ak} should go through a minimum. Structural features which increase the rotational barrier should decrease ΔS_{Ak} and therefore increase p_{Ak} resulting in an increase in melting point. Quarternary carbon atoms are an example of such a structural feature,

The results are in agreement with the assumption that packing effects in compounds with unbranched alkyl groups having 4–24 carbon atoms is constant, and that in such compounds the packing is determined by the alkyl groups. It is well known that for many straight chain substituted alkanes, the T_{m} values when plotted against n_C lie on two different curves. The two curves become closer as n_C increases, converging on a constant T_{m} value when n_C becomes very large. In the range of n_C studied in this work, no such convergence was observed. For X = H, CHO, and CO_2H, the even n_C compounds melt higher while for X = Cl or Br, the odd n_C compounds melt higher. This is understood if we compare the size (υ value and polarizability (α value) of the Me, Cl, and Br groups. The values of X, υ, and α are: Me, 0.52, 0.046; Cl, 0.55, 0.050; Br, 0.65, 0.079. Thus, the Me and Cl groups are comparable in size and polarizability, and the Br group is not much different. Then to a first approximation an odd straight chain chloro or bromo alkane should be equivalent to the even n_C alkane with one more carbon atom. They may therefore be expected to behave in packing like alkanes with even n_C. The fact that 1-alkanoic acids crystallize as hydrogen-bonded dimers should make no difference with regard to the odd–even alternation in melting point as our evidence suggests that the alkyl group is the major factor in packing.

It is surprising therefore, to find that in sets containing all types of alkyl groups there is no observable dependence on whether the chain is odd or even in the number of carbon atoms. In discussing the effects of other types of alkyl groups we make use of the p_{Ak} values, for which the reference alkyl group, Ak°, is unbranched with four or more carbon atoms. For Ak° the value of p_{Ak} is 0, for other alkyl groups the values are Me, 13.8, Et, 0; Pr, −17.7; iPr, 0; iBu, −22.4; tBu, 29.7; and neoPe, 96.2 or −14. The values for Et and iPr, both of which have a longest chain of two carbon atoms, are the same. The values of the Pr and iBu groups, both of which have a longest chain of three carbon atoms, are not significantly different. The presence of a quarternary carbon atom in tBu which has a longest chain of two carbon atoms or in some neoPe derivatives where the neoPe group has a longest chain of three seems to cause an increase in p_{Ak} as expected. Further generalization requires the determination of p_{Ak} values for many more alkyl groups.

h. *Substituent Effects*

The packing of alkyl amides with less than ten carbon atoms in the longest chain of the alkyl group is dominated by the carboxamido group. This is shown by the deviation of the carboxamides with Ak equal to Me, Et, Pr, iPr, Bu, iBu, Pe, Hp, and No. Though there are eight deviant T_m values for alkanes (MeEt, EtMe, PrH, iPrH; MeMe, EtH; iPrMe, tBuH) only three compounds with very low melting points are involved (EtH, 89.85; PrH, 85.45; tBuH, 113.75). If these melting points are reliable then the model breaks down at very low melting points. There may be a tendency for compounds with alkyl groups having a longest chain shorter than four carbon atoms and no quarternary carbon atom to deviate when the substituent is a planar π-bonded group. $AkCH{=}CH_2$ with Ak equal to Me, Et, or iPr; MeX with X equal to CO_2H or NO_2, EtBz, $PrCO_2H$, and iPrPh show deviations from the model but it should be noted that many other compounds in this category do not. Many more functional groups must be studied before conclusions can be drawn.

These results establish the applicability of the IMF equation to the quantitative description of substituted alkane melting points. The model can be extended by the determination of P_{Ak} values for additional alkyl groups. When enough values are available it should be possible to determine the variation of P_{Ak} with alkyl group structure. It may also be necessary to determine packing parameters P_x for substituents such as $CONH_2$ that dominate the packing contribution.

9. Arenes

a. *Introduction*

We have studied two types of arene both of which contain only sp^2 hybridized carbon atoms:

1. Arenes consisting only of six membered rings (6-ArH).
2. Those which may contain one or more five membered rings in addition to their six membered rings (5/6-ArH).

In order to apply the IMF equation to these compounds it is necessary to consider what intermolecular forces and steric effects are present in these compounds.

b. Hydrogen Bonding

The only hydrogen bonding that can occur involves weak $(C^{sp^2}H\text{-}\pi)$[41–44] hydrogen bonds that result from the interaction of the quadrupole moment of one aromatic ring with an H atom bonded to a C atom of another aromatic ring. Three types of H atom are extant in arenes (Figure 2). The first type is that in which the C atom to which the H is bonded is itself bonded to two C atoms that have only C—C bonds and is designated H_1,

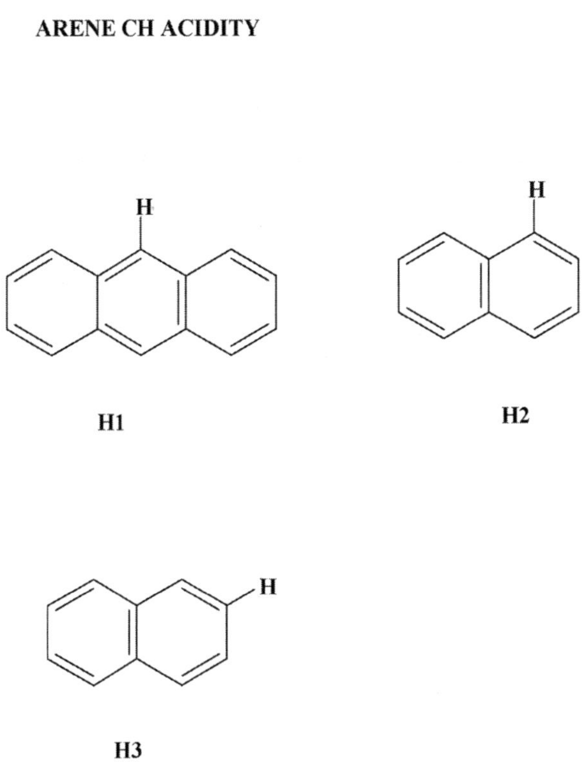

Figure 2. The acidity of arene hydrogen atoms.

the second type has the C atom to which the H is bonded itself bonded to one C atom that has only C—C bonds and is designated H_2, the third type has the C atom to which the H is bonded itself bonded to no C atom that has only C—C bonds and is designated H_3. The three types of H atom should differ somewhat in their acidity. The parameters used to account for this hydrogen bonding are n_{H1}, n_{H2}, and n_{H3}; the effective number of H atoms of each type in the arene. This differs from the total number as only some H atoms will be capable of being centered over an aromatic ring. In Table 12 when both effective and ineffective hydrogen atoms of a given type are present the number of each is given.

c. *Dispersion Forces*

As arenes which contain only six membered rings (6-ArH) or five and six membered rings (5/6-ArH) and only sp^2 hybridized C atoms have no dipole moment and there is no charge transfer interaction there is no need for the terms in σ_l, σ_d, σ_e, or μ. The only other intermolecular forces that must be parameterized are the induced dipole–induced dipole interactions (Dispersion or London forces). This type of intermolecular force is parameterized by polarizability. Polarizability can be accounted for by the parameters n_C and n_H, the numbers of C and H atoms in the arene respectively. While there is not likely to be a dependence on n_H due to the very small polarizability of the H atom the possibility cannot be excluded until it has been tested.

As imf are distance dependent and arene molecules are closer to each other in the solid than they are in the liquid state an increase in the imf should result in an increase in the value of the melting point, thus the coefficients of the imf parameters should be positive.

d. *Steric Effects*

The steric effects encountered in arenes are of four types designated SE1. SE2, SE3, and SE4 in order of their magnitude (Figure 3); they are due to H—H repulsions.[45] The parameters which account for them are n_{SE1}, n_{SE2}, n_{SE3}, and n_{SE4} respectively, they are the number of interactions of each type in the arene. The magnitude of these effects in five and six membered ring arenes is probably different from that in six membered ring arenes. The SE1 interaction in six membered ring arenes is large enough to cause noncoplanarity in the molecule resulting in chirality. As steric effects should interfere with close packing and decrease imf they should reduce the melting point. Their coefficients should therefore be negative.

Table 10. Calculated and observed values of T_m for AkX not included in set TM6

AkX	$T_{m,obs}$	$T_{m,calc}$	Δ	*NSD*	*AkX*	$T_{m,obs}$	$T_{m,calc}$	Δ	*NSD*
$MeCCl_3$	240.15	216.38	23.77	1.328	$MeCF_3$	161.85	205.81	−43.96	2.456
MeSSMe	188.45	161.16	27.29	1.525	MeSPh	258.15	154.17	103.98	5.809
$MeNHNH_2$	220.75	240.84	−20.09	1.122					
iPrSMe	171.65	150.6	21.03	1.175					
iBuI	179.65	161.36	18.29	1.022	iBuOH	165.15	176.80	−11.65	0.651
iBuSH	194.15	152.78	41.37	2.311	iBuPh	221.65	159.76	61.89	3.458
$iBuNH_2$	188.55	191.58	−3.03	0.170	$iBuCO_2H$	235.55	211.62	23.93	1.337
tBuF	196.15	193.20	2.95	0.165					
$neoPeCO_2H$	279.65	342.17	−62.52	3.493	neoPeVi	136.65	279.36	−142.71	7.972
neoPeEt	149.35	287.59	−138.24	7.723	$neoPeCH_2OH$	213.15	341.24	−128.09	7.156
neoPeMe	173.25	292.21	−118.96	6.646					
TedCHO	297.65	305.44	−7.79	0.435	$TedCO_2H$	326.65	330.17	−3.52	0.197
$TedCO_2Me$	292.65	294.87	−2.22	0.124					
OcdPh	309.15	316.58	−7.43	0.415	OcdAc	307.65	330.71	−23.06	1.288
EcH	309.55	315.40	−5.85	0.327	EcOH	345.65	352.77	−7.12	0.398
$EcCO_2H$	355.15	387.59	−32.44	1.812	EcMe	313.15	337.63	−24.48	1.368
UncH	313.15	324.97	−11.82	0.660	UncMe	317.55	347.20	−29.65	1.656
$UncCO_2H$	353.15	397.16	−44.01	2.459	$UncCO_2Me$	326.65	361.86	−35.21	1.967
$UncCO_2Et$	322.15	361.71	−39.56	2.210					
DocH	317.25	334.54	−17.29	0.966	DocOH	344.15	371.91	−27.76	1.551
DocMe	320.45	356.77	−36.32	2.029					
TrcH	320.45	344.11	−23.66	1.322					
TecH	324.15	353.68	−29.53	1.650					
HpcH	332.15	382.39	−50.24	2.807	$HpcCO_2H$	363.55	454.58	−91.03	5.086
OccH	334.65	391.96	−57.31	3.202	OccOH	356.45	429.33	−72.88	4.072
NocH	337.65	401.53	−63.88	3.569	NocOH	355.65	438.90	−83.25	4.651
TcH	341.15	411.1	−69.95	3.908	$TcCO_2H$	365.15	489.29	−118.14	6.600

Note: T_m values are from the same sources as those in Table 7.

STERIC EFFECTS IN ARENES

1 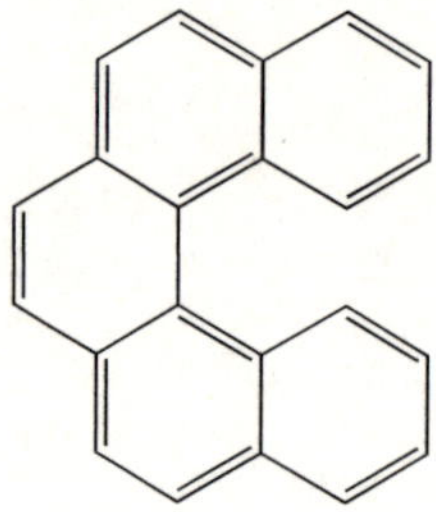 2 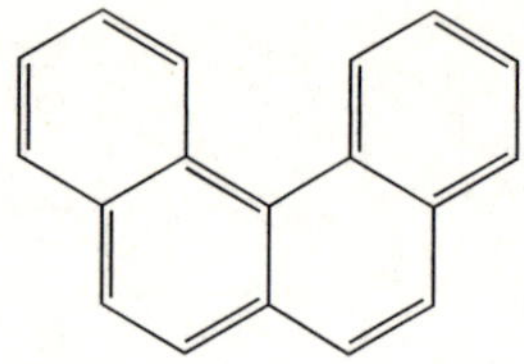

3 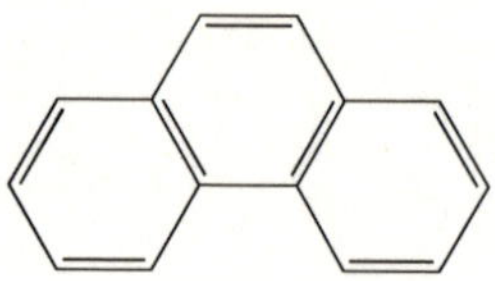 4 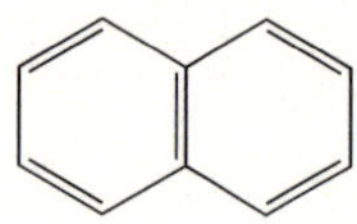

Figure 3. Steric effects in arenes

e. Packing Parameters

These include $r_{n^c/np}$, $r_{nH/nC}$, $r_{nCi/nC}$, and n_{sym}. The first three of these parameters are intended to represent the number of C atoms divided by the area of the arene. As polarizability is expected to be the greatest contributor to the imf the more C atoms in a given area the greater the bonding will be. The last represents a measure of the probability that a molecule will approach the crystal surface in a way favorable to bonding. The higher the symmetry the greater the chance of approaching in the proper arrangement to the crystal surface and bonding successfully.

The definition of $r_{n^c/np}$ requires the determination of the number of positions, n_{p}, which is given by:

$$n_p \equiv n_{\mathrm{rows(min)}} \cdot n_{\mathrm{cols(min)}} \tag{93}$$

where $n_{\mathrm{rows(min)}}$ and $n_{\mathrm{cols(min)}}$ are the minimum numbers of rows and columns respectively that include all the C atoms in the arene. Chrysene is shown as an example in Figure 4.

$r_{n^c/np}$ is then defined as:

$$r_{nC/n_p} \equiv n_C/n_p \tag{94}$$

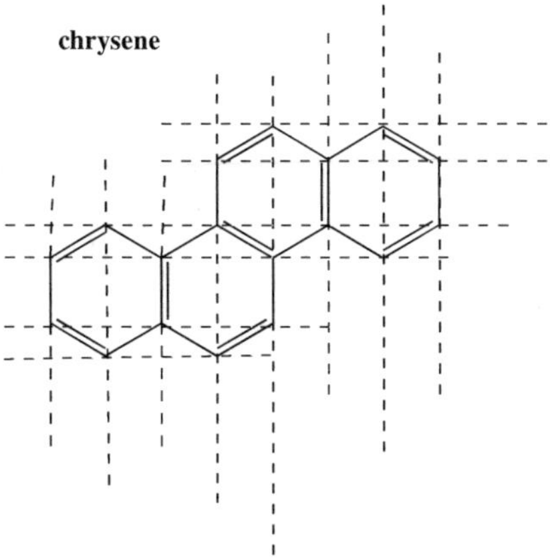

Figure 4. Rows and columns for determining n_p. The determination of p for the calculation of $r_{nC/np}$. The molecule is rotated in the plane so as to obtain the minimum value of positions p. The positions are determined by the intersection of a first set of parallel lines through the C atoms with a second set perpendicular to the first.

where n_C is the number of C atoms in the arene. The definition of r_{n^H/n^C} is:

$$r_{n_H/n_C} \equiv n_H/n_C \tag{95}$$

where n_H and n_C are the number of H and C atoms in the arene. The definition of $r_{n^{C_i}/n^C}$ is:

$$r_{n_{C_i}/n_C} = n_{C_i}/n_C \tag{96}$$

where n_{C_i} is the number of internal (nonperipheral) C atoms and n_c is the number of C atoms. n_{sym} is defined as the number of symmetry elements (planes and/or centers) in the molecule.

The larger the value of $r_{n^C/np}$ the greater the value of T_m, thus $P_{n^C/np}$ should be positive. The larger the value of r_{n^H/n^C} or $r_{n^{C_i}/n^C}$ the looser the packing, thus these parameters should have negative coefficients. The greater the value of n_{sym} the larger the melting point, then its coefficient will be positive. The correlation equation is:

$$T_m = I_C n_C + I_H n_H + H_1 n_{H1} + H_2 n_{H2} + H_3 n_{H3} + S_1 n_{SE1} + S_2 n_{SE2} + S_3 n_{SE3} + S_4 n_{SE4} + P_{r_{nH/nC}} r_{n_H/n_C} + P_{r_{nCi/nC}} r_{n_{C_i}/n_C} + P_{r_{nC/np}} r_{n_C/n_p} + P_{sym} n_{sym} + B^o \tag{97}$$

The T_m values for the six membered ring arenes (set TM7) and for the five and six membered ring arenes (set TM8) are set forth in Table 11, the values

Table 11. Melting points of arenes in sets TM7 and TM8

Arene	T_m	*Arene*	T_m
Six membered ring arenes			
Benzene **1**	279	Naphthalene **2**	353
Anthracene **3**	491	Phenanthrene **4**	373
Pyrene **5**	429	Chrysene **6**	527
Naphthacene **7**	630	Benz[a]anthracene **8**	432
Benzo[c]phenanthrene **9**	441	Triphenylene^x **10**	470
Benz[a]pyrene **11**	452	Benz[e]pyrene **12**	452
Dibenzo[de,kl]anthracene (perylene) **13**	552	Benzo[b]chrysene **14**	567
Dibenzo[a,h]anthracene **15**	539	Pentaphene **16**	530
Picene **17**	640	Coronene **18**	712
Dibenzo[fg,op]naphthalene **19**	614	Dibenzo[cd,lm]perylene **20**	648
Dibenzo[a,c]naphthacene **21**	539	Dibenzo[a,j]chrysene **22**	488
Naphtho[2,3-h]pentaphene **23**	665	Tetrabenzo[a,c,f,h]anthracene **24**	702
Ovalene **25**	746	Dibenzo[a,j]coronene **26**	771
Tetrabenzo[a,c,j,l]tetracene **27**	727	Tribenzo[de,h,kl]naphtho[1,2,3,4-rst]pentaphene^x **28**	733
Dibenzo[jk,a_1b_1]octacene **29**	833	Hexabenzo[a,cd,f,j,lm,o]perylene **30**	755
Tetrabenzo[de,no,st,c_1d_1] heptacene **31**	843	Tetrabenzo[fg,lm,uv,a1b1]-heptacene **32**	758
Dibenzo[o,s]picene^x **33**	549	Dibenz[a,j]anthracene **34**	541
Benzo[a]naphthacene **35**	537	Benzo[b]triphenylene **36**	478
Dibenzo[c,g]phenanthrene **37**	451	Dibenzo[de,mn]naphthacene **38**	535
Naphtho[1,2-a]anthracene **39**	411	Benzo[c]pentaphene **40**	555
Benzo[h]naphtho[2,1-a] anthracene **41**	658	Dibenzo[c,m]pentaphene **42**	672
Naphtho[1,2-g]chrysene **43**	461	Benzo[h]pentaphene **44**	513
Dinaphtho[2,3-a;2′,3′-h] anthracene **45**	681	Dinaphtho[1,2-b;1′,2′-k]chrysene **46**	773
Benzo[h]picene **47**	668	Dinaphtho[2,3-c;2′,3′-g]phenanthrene **48**	567
Dibenzo[a,j]naphthacene **49**	618	Dibenzo[a,l]naphthacene **50**	527
Tribenzo[a,c,j]naphthacene **51**	551	Hexaphene **52**	598
Heptaphene **53**	747	Benzo[ij]heptaphene **54**	683
Benzo[a]perylene **55**	389	Benzo[b]perylene **56**	474
Dibenzo[a,j]perylene **57**	519	Dibenzo[a,lm]perylene **58**	451
Dibenzo[a,f]perylene **59**	511	Dibenzo[fg,qr]pentacene **60**	616
Dibenzo[fg,ij]pentaphene **61**	604	Benzo[ghi]perylene **62**	546

(continued)

Table 11. Continued

Arene	T_m	*Arene*	T_m
Six membered ring arenes (continued)			
Dibenzo[b,pqr]perylene **63**	561	Tribenzo[b,k,pqr]perylene **64**	664
Tribenzo[b,m,pqr]perylene **65**	662	Tribenzo[e,k,pqr]perylene **66**	645
Naphtho[2,3-pqr]perylene **67**	542	Anthraceno[2,3-pqr]perylene **68**	629
Benzo[c]naphtho[2,3-pqr]perylene **69**	517	Dibenzo[b,n]naphtho[2,3-pqr]perylene **70**	597
Benzo[a]coronene **71**	566	Dibenzo[a,g]coronene **72**	601
Tribenzo[a,d,g]coronene **73**	624	Naphtho[2,3-a]coronene **74**	626
Dibenzo[bc,ef]coronene **75**	725	Dibenzo[fg,ij]phenanthro[2,1,10,9,8,7-pqrstuv]-pentaphene **76**	706
Dibenzo[bc,ij]diphenanthro[4,9,10-nml;8,9,10-efg]-pyrene **77**	655	Dibenzo[def,p]chrysene **78**	498
Naphtho[1,2,3,4-def]chrysene **79**	507	Dibenzo[b,def]chrysene **80**	581
Benzo[rst]pentaphene **81**	553	Dibenzo[h,rst]pentaphene **82**	594
Dibenzo[b,e,h]pentaphene **83**	511	Naphtho[2,1,8-qra]naphthacene **84**	533
Naphtho[1,2-cd]pyrene **85**	490	Naphtho[2,3-cd]pyrene **86**	546
Benzo[a]naphtho[2,3-e]pyrene **87**	536	Benzo[a]naphtho[2,3-h]pyrene **88**	587
Benzo[a]naphtho[2,3 -jk]pyrene **89**	612	Benzo[cd]naphtho[2,3-l]pyrene **90**	565
Dinaphtho[2,3-a;2′,3′-e]pyrene **91**	574	Dinaphtho[2,3-a;2′,3′-h]pyrene **92**	678
Dinaphtho[2,3-cd;2′,3′-jk]pyrene **93**	733	Dinaphtho[2,3-cd;2′,3′-l]pyrene **94**	693
Benzo[a]dinaphtho[2,3-e;2,3-jk]pyrene **95**	667	Dibenzo[au,de]pentacene **96**	558
Naphtho[1,8-auv]pentacene **97**	622	Phenanthreno[2,3-bcd]pyrene **98**	606
Benzo[a]phenanthreno[9,10-hi]naphthacene **99**	563	Tetrabenzo[a,c,hi,qr]pentacene **100**	672
Dibenzo[de,yz]hexacene **101**	618	Dibenzo[pq,wx]hexacene **102**	646
Naphtho[1,8,7-auv]hexacene **103**	658	Dibenzo[hi,yz]heptacene **104**	680
Dinaphtho[1,8,7-ac$_1$b$_1$;7,8,1-nop]heptacene[x] **105**	735	Dibenzo[def,mno]chrysene **106**	534
Pyranthrene **107**	646	Dibenzo[b,e]naphtho[1,8,7-ijk]pyrene **108**	534

(continued)

Table 11. Continued

Arene	T_m	*Arene*	T_m
Six membered ring arenes (continued)			
Tetrabenzo[fg,k,qr,ua]pentacene **109**	607	Tribenzo[de,ghi,kl]picene **110**	784
Benzo[a]naphtheno[c,d]perylene **111**	622	Naphtho[1,8,7-arq]perylene **112**	584
Pyreno[1,10,9-lmn]pyrene **113**	669	Dibenzo[ij,rst]phenanthro[7,8,9,10-defg]pentaphene **114**	803
Dinaphtho[2,3,4-rq;1′,2′,3′-hi]perylene **115**	783	Dinaphtho[1,2,3-rq;1′,2′,3′-hi]perylene **116**	751
Tetrabenzo[b,hi,k,qr]perylene **117**	617	Dinaphtho[1,8,7-rq;1′,8′,7′-hi]perylene **118**	623
Dinaphtho[2,1,8,7-arq;2′,1′,8′,7′-ghi]coronene **119**	761	Dibenzo[jk,uv]dinaphtho2,1,8,7-defg;2′,1′,8′,7′-opqr]pentacene **120**	612
Anthraceno[2,1,9,8-ghijk]pyreno[4′,5′,6′-pqa]naphthacene **121**	632	Diphenthro[5,6,7-qr;2′,3′,4′-hi]perylene **122**	648
Diphenanthro[5,6,7-qr;1,2,3-hi]perylene **123**	714	*Benzo[o]picene* **124**	718
Tribenz[a,c,h]anthracene **125**	499	*Tetrabenzo[b,g,k,p]chrysene* **126**	613
Tetrabenzo[a,d,j,m]coronene ***127***	670	*Naphtho[1,2-a]picene* **128**	563
Tetrabenzo[de,hi,mn,qr]naphthacene ***129***	579	*Tetrabenzo[de,hi,op,st]pentacene* **130**	607
Tetrabenzo[a,cd,j,lm]perylene **131**	615		
Five and six membered ring arenes			
Fluoranthene **151**	383	Benz[a]anthrylene **152**	418
Indeno[1,2,3-fg]naphthacene **153**	418	Benz[e]acephenanthrylene **154**	441
Benzo[h]fluoranthene **155**	438	Benzo[k]fluoranthene **156**	490
Dibenzo[b,k]fluoranthene **157**	509	Dibenzo[j,l]fluoranthene **158**	527
Naphtho[2,3-a]fluoranthene **159**	452	Dibenz[e,k]anthrylene **160**	503
Naphtho[2,3-j]fluoranthene **161**	500	Naphtho[2,3-k]fluoranthene **162**	525
Benzo[a]naphtho[2,3-k]fluoranthene **163**	593	Indeno[1,2,3-cd]pyrene **164**	436

(continued)

Table 11. Continued

Arene	T_m	*Arene*	T_m
Benzo[a]indeno[1,2,3-lm]pyrene **165**	536	Indeno[1,2,3-mn]naphtho[2,1,7,8-hij]tetracene **166**	566
o-Benzo[i,l]indeno[1,2,3-gf]naphtho[1,7,8-bcd]pyrene **167**	607	Benzo[mno]fluoranthene **168**	422
Dibenzo[e,mno]fluoranthene **169**	421	Indeno[1,2,3-cd]fluoranthene **170**	535
Diindeno[1,2,3-de;1′,2′,3′-mn]anthracene **171**	552	Dibenzo[a,f]indeno[1,2,3-cd]fluoranthene **172**	738
Diindeno[1,2,3-de;1′,2′,3′-kl]-anthracene **173**	579	Diindeno[1,2,3-de;1,2,3-op] naphthacene **174**	523
Diindeno[1,2,3-fg;1,2,3-mn]pyrene **175**	646	Benzo[ghi]diindeno[1,2,3-cd;1,2,3-lm]preylene **176**	715
Naphtho[1,8-j]fluoranthene **177**	506	Naphtho[1,8-k]fluoranthene **178**	588
Dinaphtho[1,8-b;1,8-g]naphthalene **179**	642	Dinaphtho[1,8-b;1,8-g]anthracene **180**	600
Dinaphtho[1,8-b;1,8-i]anthracene **181**	733	Diindeno[1,2,3-jk;1′,2′,3′-mn]phenanthene **182**	562
Indeno[1,2,3-mn]naphtho[1,8-b]pyrene **183**	623	Dinaphtho[1,8-j;1,8-l]fluoranthene **184**	660
Tribenzo[a,f,j]fluoranthene **185**	571	Naphtho[1,2,3,4-ghi]fluoranthene **186**	453
Naphtho[1,2-a]fluoranthene **187**	454	Benzo[de]indeno[1,2,-g]naphthalene **188**	484
Dibenz[a,e]aceanthrylene **189**	505	Acenaphthylene **190**	366

Notes: T_m were taken from refs. 32–35 and 51.
Compounds in italics were not included in the correlation
^{x}Excluded from the correlation.

of the parameters used in the correlations are given in Tables 12 and 13 for sets TM7 and TM8 respectively.

Of the 122 six membered ring arenes in set TM7 seven did not fit the model, they are indicated in Table 11. The cause of these deviations is uncertain. On exclusion of the variables whose coefficients were shown by a Student t test not to be significant at the 90% confidence level (CL) the

Table 12. Melting points and parameter values for 6-ArH

Arene	T_m	ΔT_m	n_C	n_H	n_{H1}	n_{H2}	n_{H3}	n_{SE1}	n_{SE2}	n_{SE3}	n_{SE4}	$r_{nC/np}$	$r_{nH/nC}$	r_{nCi}/n^C	n_{sym}
1	279	1.3	6	6	0	0	6	0	0	0	0	0.50	1.00	0	6
2	353	2.5	10	8	0	4	4	0	0	0	2	0.50	0.80	0	2
3	491	23.2	14	10	2	4	4	0	0	0	4	0.50	0.71	0	2
4	373	−33.5	14	10	0	6	4	0	0	1	2	0.44	0.71	0	1
5	429	−35.6	16	10	0	5	2	0	0	0	4	0.40	0.63	0.13	3
6	527	28.5	18	12	0	8	4	0	0	2	2	0.45	0.67	0	1
7	630	75.1	18	12	4	4	4	0	0	0	6	0.50	0.67	0	2
8	432	7.1	18	12	1+1	6	4	0	0	1	4	0.38	0.67	0	0
9	441	37.4	18	12	0	6+2	4	0	1	0	3	0.49	0.67	0	1
10[x]	470		18	12	0	0	3	0	0	6	6	0.38	0.67	0	6
11	452	−18.6	20	12	1	8	3	0	0	1	3	0.42	0.60	0.10	0
12	452	−42.7	20	12	0	8	4	0	0	2	2	0.42	0.60	0.10	1
13	552	−45.4	20	12	0	8	4	0	0	2	2	0.50	0.60	0.10	3
14	567	73.5	22	14	1+1	7+1	4	0	0	2	4	0.37	0.64	0	0
15	539	28.7	22	14	2+2	6	4	0	0	2	4	0.37	0.64	0	1
16	530	−39.8	22	14	2+2	6	4	0	0	1	6	0.50	0.37	0	1
17	640	55.5	22	14	0	10	4	0	0	3	2	0.46	0.64	0	1
18	712	0.2	24	12	0	12	0	0	0	0	6	0.43	0.50	0.25	6
19	614	−4.9	24	14	0	8	6	0	0	4	0	0.40	0.58	0.08	3
20	648	−24.8	26	14	0	12	2	0	0	2	4	0.43	0.54	0.15	3
21	539	−40.9	26	16	4	6	6	0	0	3	4	0.33	0.62	0	1
22	488	−65.0	26	16	0	4	8	0	2	2	0	0.46	0.62	0	3
23	665	16.2	30	18	6	6	6	0	0	3	6	0.28	0.60	0	3
24	702	−2.0	30	18	2	4	8	0	0	6	0	0.42	0.60	0	3
25	746	9.3	32	14	2	12	0	0	0	0	8	0.44	0.44	0.31	3
26	771	58.8	32	16	0	12	4	0	0	4	2	0.36	0.50	0.19	3

27	727	−16.3	34	20	4	4+4	8	0	0	2	6	0.39	0.59	0	3
28[x]	733		38	20	0	12	8	0	4	0	2	0.38	0.53	0.11	3
29	833	−27.1	40	22	8	8	6	0	0	2	4	0.31	0.55	0.05	3
30	755	33.9	42	22	0	12	10	2	0	4	0	0.39	0.52	0.10	3
31	843	−34.9	42	22	6	12	4	0	0	4	8	0.35	0.52	0.10	2
32	758	−78.1	42	22	2	4+8	8	0	0	8	0	0.35	0.52	0.10	3
33[x]	549		30	18	0	8+2	8	0	0	2	4	0.42	0.53	0	1
34	541	3.4	22	14	1+1	6+2	4	0	0	2	4	0.41	0.64	0	1
35	537	43.5	22	14	3+1	5+1	4	0	0	1	6	0.37	0.64	0	0
36	478	−17.3	22	14	0+2	6	6	0	0	3	2	0.34	0.64	0	1
37	451	−51.4	22	14	0	8+2	4	1	0	0	4	0.46	0.64	0	1
38[x]	535		24	14	0+2	4+4	4	0	0	2	4	0.40	0.58	0.08	1
39	411	−57.1	22	14	1+1	7+1	1	0	1	0	5	0.43	0.64	0	0
40	555	−6.1	26	16	1+3	8	1	0	0	2	6	0.37	0.62	0	0
41[x]	658		26	16	0+2	10	1	0	0	3	4	0.37	0.62	0	0
42	672	6.8	30	18	0+4	10	1	0	0	3	6	0.38	0.60	0	1
43	461	−10.2	26	16	0+2	7+1	6	0	1	2	3	0.33	0.62	0	0
44	513	−27.7	26	16	0+4	6	6	0	0	3	4	0.31	0.62	0	1
45	681	20.5	30	18	2+4	9+2	4	0	0	2	8	0.40	0.60	0	1
46	773	−73.9	34	20	0+4	12	4	0	0	4	6	0.31	0.59	0	1
47[x]	668		26	16	1+1	8+2	4	0	0	3	4	0.37	0.62	0	0
48	567	−8.2	30	18	2+2	8+2	4	2[a]	0	0	6	0.47	0.60	0	1
49	618	77.3	26	16	2+2	8	4	0	0	2	6	0.31	0.62	0	1
50	527	−51.4	26	16	2+2	6+2	4	0	0	2	6	0.39	0.62	0	1
51	551	−62.8	30	18	1+3	7+1	3	0	0	4	4	0.34	0.60	0	0
52	598	35.4	26	16	4+2	6	4	0	0	1	8	0.31	0.62	0	0
53[x]	747		30	18	4+2	6	4	0	0	1	10	0.31	0.60	0	1
54	683	42.4	34	20	4+4	6	6	0	0	3	8	0.27	0.59	0	1

(continued)

Table 12. Continued

Arene	T_m	ΔT_m	n_C	n_H	n_{H1}	n_{H2}	n_{H3}	n_{SE1}	n_{SE2}	n_{SE3}	n_{SE4}	$r_{nC/np}$	$r_{nH/nC}$	r_{nci}/n^c	n_{sym}
55	389	−53.0	24	14	1+0	6+2	5	0	1	1	3	0.43	0.58	0.08	0
56	474	−65.6	24	14	0+1	8	5	0	0	3	2	0.40	0.58	0.08	0
57	519	28.5	28	16	2+0	4+4	6	0	2	0	4	0.39	0.57	0.07	1
58	451	−73.1	28	16	1+1	7+1	6	0	1	2	3	0.40	0.57	0.07	0
59	511	20.5	28	16	2+0	4+4	6	0	2	0	4	0.40	0.57	0.07	1
60	616	0.1	28	16	0+2	8	6	0	0	4	2	0.40	0.57	0.07	1
61	604	−11.9	28	16	0+2	8	6	0	0	4	2	0.40	0.57	0.07	1
62	546	6.6	22	12	0	10	2	0	0	1	4	0.46	0.55	0.18	1
63	561	−2.9	26	14	0	10	4	0	0	3	2	0.43	0.54	0.15	0
64	664	69.7	30	16	0+1	10	5	0	0	4	1	0.36	0.53	0.13	0
65	662	6.4	30	16	0	10	6	0	0	5	0	0.42	0.53	0.13	1
66	645	19.3	30	16	0+2	10	4	0	0	3	4	0.36	0.53	0.13	1
67	542	−43.3	26	14	0	10	4	0	0	3	2	0.41	0.54	0.15	1
68	629	−6.6	30	16	0+2	10	4	0	0	3	4	0.38	0.53	0.13	1
69	517	−17.2	30	16	0	8+2	6	0	1	3	1	0.38	0.53	0.13	0
70	597	47.6	34	18	0	6+4	8	0	2	3	0	0.38	0.53	0.12	1
71	566	−44.1	28	14	0	12	2	0	0	2	4	0.40	0.50	0.21	1
72	601	−58.4	32	16	0	12	4	0	0	4	2	0.38	0.50	0.19	1
73	624	35.4	36	18	0	8+4	6	0	2	2	2	0.40	0.50	0.17	1
74	626	−33.4	32	15	0+2	12	2	0	0	0	8	0.38	0.50	0.19	1
75	725	70.9	30	14	2	10	2	0	0	1	6	0.47	0.47	0.27	1
76	706	22.1	34	16	0	12	4	0	0	4	2	0.40	0.47	0.24	1
77	655	−16.9	42	20	0	8+4	8	0	2	4	0	0.42	0.48	0.19	1
78	498	21.0	24	14	1	6+2	5	0	1	1	3	0.43	0.58	0.08	0
79	507	−7.7	24	14	0+1	8	5	0	0	3	2	0.38	0.58	0.08	0
80	581	10.3	24	14	2	8	4	0	0	2	4	0.40	0.58	0.08	1

81	553	−17.7	24	14	2	8	4	0	0	2	4	0.40	0.58	0.08	1
82	594	13.1	28	16	0+2	8	6	0	0	4	2	0.33	0.57	0.07	1
83	511	25.5	28	16	2	4+4	6	0	2	0	4	0.39	0.57	0.07	1
84	533	6.8	24	14	0+2	8	4	0	0	2	4	0.34	0.58	0.08	1
85	490	−6.6	24	14	1	8+2	3	0	1	0	5	0.44	0.58	0.08	0
86	546	6.7	24	14	2+1	8	3	0	0	1	6	0.40	0.58	0.08	0
87	536	−30.8	28	16	1+2	8	5	0	0	3	4	0.35	0.57	0.07	0
88	587	6.1	28	16	0+2	8	6	0	0	4	2	0.33	0.57	0.07	1
89	612	10.6	28	16	3+1	8	4	0	0	2	6	0.39	0.57	0.07	0
90	565	−12.5	28	16	3+1	8	4	0	0	2	6	0.33	0.57	0.07	0
91	574	−22.9	32	18	1+4	8	5	0	0	3	6	0.29	0.56	0.06	0
92	678	5.9	32	18	0+4	8	6	0	0	4	4	0.33	0.56	0.06	2
93	733	38.1	32	18	4+2	8	4	0	0	2	8	0.38	0.56	0.06	1
94	693	37.6	32	18	4+2	6+2	4	0	0	2	8	0.33	0.56	0.06	1
95	667	3.0	36	20	2+4	6+2	6	0	0	4	6	0.28	0.56	0.06	1
96	558	−32.6	28	16	2+2	8	4	0	0	2	6	0.32	0.57	0.07	1
97	622	13.3	28	16	4+1	8	3	0	0	1	8	0.39	0.57	0.07	0
98	606	9.6	28	16	1+2	10	3	0	0	2	6	0.38	0.57	0.07	0
99	563	12.7	32	18	0	6+4	8	0	2	3	0	0.40	0.56	0.06	1
100	672	−27.2	36	20	0+2	6+4	8	0	0	7	0	0.38	0.56	0.06	1
101	618	−27.4	32	18	4+2	6+2	4	0	0	2	8	0.31	0.56	0.06	1
102	646	0.6	32	18	2+2	8	6	0	0	4	4	0.31	0.56	0.06	1
103	658	−20.2	32	18	6+1	8	3	0	0	1	10	0.38	0.56	0.06	0
104	680	−8.6	36	20	2+4	8	6	0	0	4	6	0.30	0.56	0.06	1
105^{x}	735		42	22	6+2	12	2	0	0	2	12	0.31	0.52	0.10	1
106	534	−5.4	22	12	2	8	2	0	0	0	6	0.46	0.55	0.18	1
107	646	20.3	30	16	2+2	8	4	0	0	2	6	0.36	0.53	0.13	1
108	534	−0.2	30	16	2+1	6+2	5	0	1	1	5	0.38	0.53	0.13	0

(*continued*)

Table 12. Continued

Arene	T_m	ΔT_m	n_C	n_H	n_{H1}	n_{H2}	n_{H3}	n_{SE1}	n_{SE2}	n_{SE3}	n_{SE4}	$r_{nC/np}$	$r_{nH/nC}$	r_{nci}/n^c	n_{sym}
109	607	33.0	34	18	0	8+4	6	0	2	2	2	0.40	0.53	0.12	1
110	784	42.4	30	16	0	12	4	0	0	4	2	0.50	0.53	0.13	2
111	622	48.0	34	18	0	8+4	6	0	2	2	2	0.40	0.53	0.12	1
112	584	20.1	26	14	0+1	10	3	0	0	2	4	0.43	0.54	0.15	0
113	669	24.0	28	14	0	12	2	0	0	2	4	0.47	0.50	0.21	1
114	803	59.4	36	18	0	12	6	0	0	6	0	0.43	0.50	0.17	1
115	783	39.4	34	18	2	12	4	0	0	4	4	0.43	0.53	0.12	1
116	751	7.4	34	18	2	12	4	0	0	4	4	0.43	0.53	0.12	1
117	617	8.1	34	18	0	8+4	6	0	2	2	2	0.47	0.53	0.12	1
118	623	−66.3	32	16	0+2	12	2	0	0	2	6	0.44	0.50	0.19	1
119	761	27.3	36	16	2	12	2	0	0	2	6	0.43	0.44	0.28	1
120	612	−41.0	38	18	0	12+4	2	0	2	0	6	0.43	0.47	0.21	1
121	632	−39.6	38	18	1+1	13+1	2	0	1	1	7	0.39	0.47	0.21	0
122	648	−44.2	42	22	2	10+6	4	0	2	2	6	0.38	0.52	0.10	1
123	714	7.5	42	22	2	12+4	4	0	2	2	6	0.35	0.52	0.10	1
124[n]	718		26	16	0	4+8	4	0	0	4	2	0.46	0.62	0	1
125[n]	499	27	26	16	0+2	5+3	6	0	0	4	1	0.36	0.62	0	0
126[n]	613	99	34	20	0+4	4+4	8	0	2	2	4	0.28	0.59	0	1
127[n]	670	11	40	20	0	0+12	8	0	2	4	0	0.48	0.50	0.15	2
128[n]	563	72	30	18	1+1	2+10	4	0	1	2	5	0.31	0.60	0	0
129[n]	579	18	30	16	0	8+4	4	0	2	0	4	0.42	0.53	0.13	2
130[n]	607	68	34	18	0	4+8	6	0	2	2	2	0.39	0.53	0.12	1
131[n]	615	−6	32	16	0	4+8	4	0	0	4	2	0.36	0.50	0.19	2

Notes: Arene numbers refer to Table 11.
n. Not included in the data set.
[x] Excluded from the correlation.

regression equation was:

$$T_{m,ArH} = 12.0(\pm 1.16)n_C - 67.0(\pm 13.9)n_{SE1} - 56.0(\pm 6.77)n_{SE2} + 506(\pm 93.1)r_{nC/n_p} + 31.4(\pm 3.89)n_{\text{sym}} + 8.00(\pm 2.53)n_{H1} + 6.80(\pm 2.51)n_{H2} - 618(\pm 144)r_{n_H/nC} - 543(\pm 105)r_{i/t} + 382(\pm 116) \quad (98)$$

$100R^2$, 88.61; A$100R^2$, 87.75; F, 90.80; S_{est}, 37.2; S^{o}, 0.353; N_{dp}, 115.

The correlation matrix for this equation and for Eq. 98 is given in Table 14. As the coefficients of n_{H1} and n_{H2} show no significant difference the two parameters were combined into a single parameter, n_{H12} which is the total number of H atoms of types 1 and 2 in the arene. Correlation with this equation gave the regression equation:

$$T_{m,ArH} = 12.0(\pm 1.15)n_C - 67.5(\pm 13.8)n_{SE1} - 55.3(\pm 6.73)n_{SE2} + 499(\pm 92.1)r_{nC/n_p} + 31.4(\pm 3.89)n_{\text{sym}} + 7.32(\pm 2.01)n_{H12} - 605(\pm 142)r_{n_H/nC} - 553(\pm 101)r_{i/t} + 373(\pm 115) \quad (99)$$

$100R^2$, 88.55; A$100R^2$, 87.80; F, 102.5; S_{est}, 37.1; S^{o}, 0.352; N_{dp}, 115. C_{C}, 31.2; C_{SE1}, 5.83; C_{SE2}, 4.78; C_{φ}, 12.9; C_{sym}, 2.72; C_{H12}, 6.33; $C_{\text{rnH/nC}}$, 31.4; $C_{ri/t}$, 4.78. C_{imf}, 37.5; C_{steric}, 10.6; C_{packing}, 51.9.

Packing parameters make the greatest contribution followed by imf parameters (polarizability and CH-π interactions). There is also a dependence on steric effects. As predicted, the coefficients of n_{C} and n_{12} are positive and there is no dependence on n_{H}. There is no dependence on n_{H3} indicating that the third type of hydrogen atom is not involved in CH-π interactions. There is no dependence on n_{SE3} or n_{SE4} suggesting that these steric interactions have a negligible effect on T_{m}. The coefficients of the packing parameters n_{sym} and $r_{\text{nC/np}}$ are positive as predicted while those of $r_{\text{nH/nC}}$ and $r_{\text{nCi/nC}}$ are negative, again as predicted. The arenes not included in the correlation, numbers 124–131, give calculated T_{m} values that are generally within two standard deviations of the observed values.

For the case of five and six membered rings in the molecule the best regression equation is:

$$T_m = 12.2(\pm 0.925)n_C + 9.37(\pm 4.77)n_{SE2} + 15.4(\pm 2.86)n_{SE3} + 32.2(\pm 6.37)n_{\text{sym}} + 19.5(\pm 5.13)n_{Hl} + 147(\pm 21.2) \quad (100)$$

$100R^2$, 93.28; A$100R^2$, 92.49; F, 91.62; S_{est}, 25.9; S^{o}, 0.282; N_{dp}, 39. C_{C}, 79.2; C_{SE2}, 2.02; C_{SE3}, 3.33; C_{sym}, 6.98; C_{H1}, 8.45; C_{imf}, 87.7; C_{packing}, 6.98, C_{other}, 5.35

The correlation matrix for Eq. 100 is given in Table 15. The other interactions (SE2 and SE3) may also be packing interactions rather than steric effects as they have the wrong sign for steric effects. As the hydrogen

Table 13. Melting points and parameter values for 5,6-ArH

Arene	T_m	ΔT_m	n_C	n_H	n_{H1}	n_{H2}	n_{H3}	n_{SE1}	n_{SE2}	n_{SE3}	n_{SE4}	$r_{nC/np}$	$r_{nH/nC}$	r_{nci}/n^e	n_{sym}
151	383	21.8	16	10	0	6	4	0	0	2	1	0.46	0.63	0.06	1
152	418	16.9	20	12	1	3+2	6	0	1	1	2	0.41	0.60	0.05	0
153	490	39.4	24	14	2	2+4	6	0	2	0	4	0.43	0.58	0.04	1
154	441	15.3	20	12	1	7	6	0	0	2+1	1	0.36	0.60	0.05	0
155	438	−22.6	20	12	0	8	4	0	1	1	2	0.36	0.60	0.05	0
156	490	2.6	20	12	2	6	4	0	0	2	3	0.44	0.60	0.05	1
157	509	35.1	24	14	3	6	5	0	0	2+1	3	0.36	0.58	0.04	0
158	527	−21.2	24	14	0	4+4	6	0	2	0+1	1	0.38	0.58	0.04	1
159	452	22.3	24	14	1	6+2	5	1	0	1+0	3	0.43	0.58	0.04	0
160	503	2.0	24	14	1+1	6+1	5	0	0	2+1	2	0.43	0.58	0.04	0
161	500	−16.3	24	14	1+1	2+1	4	0	1	1+0	4	0.34	0.58	0.04	0
162	525	16.4	24	14	2+2	6	4	0	0	2+0	5	0.44	0.58	0.04	1
163	593	−21.5	28	16	3+2	6	5	0	1	1+0	6	0.36	0.57	0.04	0
164	436	29.3	22	12	1+0	8+0	3	0	0	2+0	3	0.41	0.55	0.14	0
165	536	−12.6	26	14	1+0	6+2	5	0	1	1+1	2	0.36	0.54	0.12	0
166	566	−19.3	30	16	0+1	6+4	5	0	2	0+1	3	0.33	0.53	0.10	0
167	607	22.3	36	18	0+1	3+5	7	0	2+1	0+1	2	0.40	0.50	0.14	0
168	422	−8.2	18	10	0	8	2	0	0	1+0	3	0.42	0.53	0	1
169	421	44.3	22	12	1	8	3	0	0	1+1	3	0.41	0.55	0.05	0
170	535	6.0	22	12	0	8	4	0	0	4+0	0	0.44	0.55	0.09	2
171	532	−6.5	26	14	0	4+4	6	0	2	2+0	0	0.37	0.54	0.08	1
172	738	−0.4	30	16	0	0+8	8	0	4	0	0	0.33	0.53	0.07	2
173	579	−33.5	26	14	0	4+4	6	0	2	2+0	0	0.32	0.54	0.08	1
174	523	33.1	30	16	0+1	3+5	7	0	3	1+0	0	0.33	0.53	0.07	0
175	646	−25.0	28	14	2+0	8+0	4	0	0	4+0	2	0.39	0.50	0.14	1

176	715	−20.8	34	16	2+0	10+0	4	0	0	4+1	2	0.41	0.47	0.12	1
177	506	20.8	26	14	0	8+2	4	1	0	2+0	2	0.48	0.54	0.08	1
178	588	1.8	26	14	0+2	8+0	4	0	0	4+0	2	0.42	0.54	0.08	2
179	642	−3.4	30	16	0+4	8+0	4	0	0	4+0	4	0.46	0.53	0.07	2
180	600	12.4	34	18	0+2	6+4	6	1	2	0	4	0.34	0.53	0.06	1
181	733	−6.6	34	18	2+4	8+0	4	0	0	4+0	6	0.45	0.53	0.06	2
182	562	−19.8	26	14	0	6+2	6	1	0	2+1	0	0.45	0.54	0.08	1
183	623	−41.8	32	16	1+0	6+4	5	1	1	1+0	3	0.40	0.50	0.13	0
184	660	22.5	36	18	0	6+6	6	3	0	0	3	0.36	0.50	0.08	3
185	571	−34.4	28	16	2	6+2	6	1	1	0	4	0.39	0.57	0.04	0
186	453	8.1	22	12	0	8	4	0	0	1+2	1	0.45	0.55	0.14	0
187	454	25.5	24	14	0+1	8	5	0	1	1+1	2	0.33	0.58	0.04	0
188	484	−43.1	20	12	1+1	6+0	4	0	0	1+1	3	0.42	0.60	0.05	0
189	505		24	12	0+2	0+6	6	0	0	2+1	0	0.27	0.58	0.04	0
190	366	−40.7	12	8	0	6+0	2	0	0	0	3	0.48	0.67	0.08	1

Table 14. Correlation matrices for Eqs. 98 and 99

n_C	$n_{H1}(n_{H12})$	n_{H2}	n_{SE1}	n_{SE2}	$r_{nC/np}$	$r_{nH/nC}$	$r_{nCi/nC}$	n_{sym}	
1	0.187(**0.507**)	**0.449**	0.117	0.287	**0.400**	**0.764**	**0.383**	0.063	n_C
	1	0.163	0.036	0.129	**0.292**	0.004	0.198	0.151	n_{H1}
			(0.022)	(0.219)	(0.096)	(**0.646**)	(**0.536**)	(0.150)	(n_{H12})
		1	0.003	0.144	0.068	**0.706**	**0.707**	0.070	n_{H2}
			1	0.076	0.131	0.006	0.093	0.100	n_{SE1}
				1	0.094	0.185	0.111	0.105	n_{SE2}
					1	0.167	0.234	0.228	$r_{nC/np}$
						1	**0.753**	0.183	$r_{nH/nC}$
							1	0.101	$r_{nCi/nC}$
								1	n_{sym}

***Note*:** Values for n_{H12} are in parentheses. Values in boldface are at the 99.9% confidence level.

Table 15. Correlation matrix for Eq. 100

n_C	n_{H1}	n_{SE2}	n_{SE3}	n_{sym}	
1	0.057	0.386	0.119	0.261	n_C
	1	0.269	0.008	0.239	n_{H1}
		1	0.040	0.086	n_{SE2}
			1	0.365	n_{SE3}
				1	n_{sym}

Note: Values in boldface are at the 99.9% confidence level.

atoms involved in these steric effects are directed towards each other they are poor candidates for CH-π interactions. The coefficient of n_{SE2} is barely significant and may simply be an artifact. The coefficients of n_C and n_{H1} are positive as predicted, the magnitude of the former is not significantly different from that in Eq. 99. The coefficient of n_{sym} is positive as predicted and again is not significantly from that in Eq. 99. The lack of dependence on the packing parameters r_{nC}/nP, $r_{nH/nC}$, and $r_{nCi/nC}$ may be due to the smaller size of the data set and the smaller range of the parameters as shown in Table 16.

f. Conclusions

The following conclusions may be drawn:

1. QSPR for T_m of 6-ArH and 5,6-ArH were obtained using a variant of the imf model.
2. Steric effects of types SE1 and SE2 were significant for 6-ArH. They are not significant for 5,6-ArH.
3. CH-π interactions are significant for 6-ArH. H_3 type H atoms are probably not involved in this interaction.
4. C atom densities as measured by $r_{nC/np}$, $r_{nH/nC}$, and $r_{nCi/nC}$ are a significant packing factor in 6-ArH The range of certainly the latter two parameters and possibly the first as well coupled with the smaller data set size makes the dependence for 5,6-ArH uncertain.
5. n_H is not a significant imf parameter for 6-ArH or 5,6-ArH. This may be due to its high linearity in n_c.
6. n_{sym} is a significant packing parameter for 6-ArH and 5,6-ArH.

Table 16. Range of r_{nC}/nP, $r_{nH/nC}$, and $r_{nCi/nC}$

	$r_{nC/np}$ Minimum value	Maximum value	Range	$r_{nH/nC}$ Minimum value	Maximum value	Range	$r_{nCi/nC}$ Minimum value	Maximum value	Range
6-ArH	0.27	0.50	0.23	0.40	1.00	0.60	0	0.31	0.31
5,6-ArH	0.33	0.48	0.15	0.50	0.67	0.17	0.04	0.14	0.10
$r_{6/5,6}$			1.5			3.5			3.1

Note: $r_{6/5,6}$ is the ratio of the range of the 6-ArH to that of the 5,6-ArH.

10. Melting Points of X-C(sp^2) Compounds

Melting points of XG where G is vinyl, 1-buta-1,3-dienyl, phenyl, E-1-vinyl-2-phenyl, 4-vinyl-1-phenyl, 4-biphenyl, 2-naphthyl, 2-anthracyl, and 2- and 3-phenanthryl (set TM9, Table 17) have been correlated with a variant of the IMF equation. Most of the parameters are set forth in Table 8. The

Table 17. T_m for X-C(sp^2) compounds

ViX. X, T_m: OAc, 180; Br, 134; Cl, 119; F, 113; CCl_3, 243; Me, 88.0; CH_2Cl, 139; CH_2Br, 154; CH_2I, 174; CH_2OH, 144; CO_2H, 286; CO_2Bu, 209; CO_2Et, 202; $CONH_2$, 358; CN, 191; Vi, 164; Et, 87.9; OEt, 157; H, 104; NO_2, 218; Bu, 133; OMe, 151; CHO, 186; sBu, 120; tBu, 158; iBu, 120; Pr, 135; CH_2CN, 189.
E-2-ViVnX. X, T_m: CN, 230; CO_2H, 353; Me, 186; H, 164; Et, 171; Vi, 261.
PhX. X, T_m: H, 279; F, 232; Cl, 228; Br, 242; I, 242; CN, 260; NO_2, 279; Me, 178; Et, 178; iPr, 250; tBu, 215; OH, 316; OMe, 236; OEt, 244; NH_2, 267; NMe_2, 276; CHO, 232; Ac, 294; CO_2H, 295; $CONH_2$, 406; CO_2Me, 261; CO_2Et, 239; SO_2NH_2, 429; 2-MeVn, 246; EtCO, 292; $NHCO_2Et$, 326; NHMe, 216; CH_2CO_2H, 350; $NHNH_2$, 293; NCO, 243; NCS, 252; NHPh, 328; E—N_2Ph, 342; CH_2NH_2, 283; NHBz, 436; OBz, 343; COCl, 272; COF, 245; PrCO, 286; Bz, 321 (α), 299 (β); CH_2OH, 258; CH_2F, 238; CH_2Cl, 234; CH_2Br, 270; CH_2CN, 249; CH_2Vi, 233; CH_2Ph, 299; CF_3, 244; CCl_3, 268; Vi, 243; C_2H, 228; Ph, 344; N_3, 244; SH, 258; SPh, 247; S_2Ph, 334; SO_2Ph, 402; SOPh, 344; SO_2Cl, 287; SO_2OH, 324; SePh, 276; Se_2Ph, 335; $POCl_2$, 276; $PO(OH)_2$, 438; $AsO(OH)_2$, 436; NHAc, 387; Pr, 174; OBu, 254; iBu, 222; Bu, 185; OPr, 246; OiPr, 240; Pe, 198; sBu, 198; cHx, 280; $NHCONH_2$, 420.
4-X-Styrene. X, T_m: H, 243; Me, 235; Et, 223; Br, 278; Cl, 257; F, 239; NO_2, 302; NH_2, 297.
E-2-PhVnX. X, T_m: H, 240; Ph, 397; Br, 279; NO_2, 331; CH_2Cl, 254; CH_2OH, 306; Vi, 278; Ac, 315; CHO, 266; CO_2H, 409; CN, 295; COCl, 309; CO_2Me, 310; CO_2Et, 281; $CONH_2$, 421.
2-NhX. X, T_m: H, 354; F, 334; Cl, 334; Br, 332; I, 327; Me, 308; Et, 266; Bu, 268; tBu, 269; CH_2Cl, 322; CH_2Br, 329; CH_2OH, 335; CH_2CN, 359; Vi, 339; Ph, 377; CHO, 335; COCl, 324; Ac, 329; CO_2H, 459; CO_2Me, 350; CO_2Et, 305; $CONH_2$, 468; CN, 339; NH_2, 386; NMe_2, 327; NHAc, 407; NO_2, 352; OH, 397; OMe, 347; OEt, 311; OPr, 314; OAc, 343; SH, 354; SCN, 308.
4-PhPnX. X, T_m: H, 344; F, 348; Cl, 351; Br, 364; I, 387; Me, 323; CH_2Ph, 358; Vi, 392; Ac, 394; CO_2H, 501; CN, 361; NH_2, 327; NHAc, 445; NO_2, 387; OH, 438; OMe, 363.
2-X-Anth. X, T_m: H, 489; Cl, 492; Me, 482; Vi, 398; Ac, 464; NH_2, 513; OH, 526.
2-XPnth. X, T_m: H, 373; Ac, 418; CO_2H, 532; CN, 382; NH_2, 358; OH, 441.
3-XPnth. X, T_m: H, 373; Me, 338; Ac, 355; CO_2H, 543; CN, 375; NH_2, 416 (α), 361 (β); NO2, 446; OH, 395.

Note: T_m were taken from refs. 32–36.

packing parameter P_G was set equal to 0 for G equal to phenyl. Packing parameters for the remaining groups G were determined by the same method used for the melting points of substituted alkanes described above. Best results were obtained on the introduction of the hydrogen bonding parameters hb_{OO} and hb_{ON}. These parameters represent an OH...O and an NH...O hydrogen bond respectively in the crystal. The correlation equation is:

$$T_{mXG} = L\sigma_{lX} + D\sigma_{dX} + R\sigma_{eX} + M\mu X + A\alpha_X + H_1 n_{HX} + H_2 n_{nX} + B_{hbOO} n_{hbOO} + B_{hbON} n_{hbON} + S_1 \upsilon_{1X} + S_2 \upsilon_{2X} + S_3 \upsilon_{3X} + B_C n_{CG} + B_p p_G + B^o \quad (101)$$

The regression equation is:

$$T_{mXG} = -135(\pm 19.5)\sigma_{lX} + 58.6(\pm 8.98)\sigma_{dX} - 152(\pm 48.0)\sigma_{eX} - 12.9(\pm 3.56)\mu_X + 180(\pm 29.1)\alpha_X + 34.6(\pm 4.81)n_{HX} + 4.71(\pm 1.81)n_X + 62.7(\pm 8.57)hb_{OO} + 91.5(\pm 11.5)hb_{ON} + 1.00(\pm 0.0253)p_G + 204(\pm 5.01) \quad (102)$$

$100R^2$, 93.74; A$100R^2$, 93.40; F, 249.9; S_{est}, 24.0; S^o, 0.258, N_{dp}, 178. $C_{\sigma l}$, 29.9; $C_{\sigma d}$, 12.9; $C_{\sigma e}$, 3.37; C_{μ}, 2.85; C_{α}, 7.95; C_{nH}, 7.65; C_{nn}, 1.04; C_{hbOO}, 13.9; C_{hbON}, 20.2; C_{pG}, 0.221.

The correlation matrix is given in Table 18.

The nature of the packing parameter p_G was determined by correlation with the equation:

$$p_G = a_C n_C + a_s n_s + a_{rb} n_{rb} + a_{arf} n_{arf} + a_o \quad (103)$$

Table 18. Correlation matrix for Eq. 102

σ_l	σ_d	σ_e	μ	α	n_H	n_n	hb_{OO}	hb_{ON}	P_G	
1	0.117	0.065	0.761	0.127	0.061	0.408	0.045	0.028	0.083	σ_l
	1	0.133	0.356	0.030	0.241	0.272	0.030	0.012	0.065	σ_d
		1	0.178	0.374	0.235	0.206	0.084	0.131	0.024	σ_e
			1	0.077	0.190	0.569	0.035	0.392	0.037	μ
				1	0.164	0.022	0.235	0.050	0.266	α
					1	0.361	0.366	0.537	0.088	n_H
						1	0.357	0.342	0	n_n
							1	0.074	0.141	hb_{OO}
								1	0.029	hb_{ON}
									1	P_G

where: n_C is the number of C atoms in G; n_s is the number of planes of symmetry in XG. n_{rb} is the number of bonds in G about which rotation can take place; and n_{arf} is the number of angular ring fusions in G.

The regression equation is:

$$P_G = 24.0(\pm 2.54)n_C - 61.4(\pm 26.1)n_{arf} - 142(\pm 22.4) \qquad (104)$$

$100R^2$, 93.74; $A100R^2$, 92.96; F, 52.45; S_{est}, 26.9; S^o, 0.299; N_{dp}, 10. C_{nC}, 79.7; C_{narf}, 20.3. p_G is predominantly dependent on n_C and represents ii interactions.

On introducing the variables of which P_o is a function the regression equation is:

$$\begin{aligned} T_{mXG} = &-135(\pm 19.6)\sigma_{lX} + 58.6(\pm 9.06)\sigma_{dX} - 152(\pm 48.4)\sigma_{eX} \\ &- 12.9(\pm 3.58)\mu_X + 180(\pm 29.6)\alpha_X + 34.6(\pm 4.82)n_{HX} \\ &+ 4.71(\pm 1.81)n_{nX} + 62.7(\pm 8.62)hb_{OO} + 91.5(\pm 11.6)hb_{ON} \\ &+ 1.00(\pm 0.0257)p_G - 20.9(\pm 10.3)n_{arf} + 204(\pm 5.20) \qquad (105) \end{aligned}$$

$100R^2$, 93.74; $A100R^2$, 93.36; F, 225.9; S_{est}, 24.1; S^o, 0.259; n, 178. $C_{\sigma l}$, 28.5; $C_{\sigma d}$, 12.4; $C_{\sigma e}$, 3.22; C_{μ}, 2.73; C_{α}, 7.60; C_{nH}, 7.31; C_{nn}, 0.995; C_{hbOO}, 13.3; C_{hbON}, 19.3; C_{pG}, 0.222; C_{nG}, 4.41.

The correlation matrix is given in Table 19.

There is no dependence on steric effects in this model. The inclusion of the term in n_{arf} has not resulted in a significant improvement. The parameters used in this correlation are reported in Table 8.

11. Substituted Pyridines

The data sets studied are reported in Table 20. Due to the small number of data points in the data sets, hydrogen bonds are accounted for by the parameter n_{hb}, the number of hydrogen bonds formed by a molecule. Thus, carboxyl groups are expected to form two hydrogen bonds, hydroxyl groups only one.

The regression equations and correlation matrices obtained are for the 2-substituted compounds (set TM10, Table 20);

$$\begin{aligned} T_m = &-19.8(\pm 6.84)\mu + 70.6(\pm 7.36)n_{hb} - 79.6(\pm 33.4)\upsilon_2 \\ &+ 169(\pm 42.2)\sigma_l - 849(\pm 203)\sigma_e + 14.8(\pm 13.4) \qquad (106) \end{aligned}$$

$100R^2$, 91.85; $A100R^2$, 89.81; F, 33.81; S_{est}, 24.8; S^o, 0.338; N_{dp}, 21.

Table 19. Correlation matrix for Eq. 105

σ_I	σ_d	σ_e	μ	α	n_H	n_n	hb_{OO}	hb_{ON}	P_G	n_{arf}	
1	0.117	0.007	0.761	0.127	0.061	0.408	0.045	0.028	0.076	0.106	σ_I
	1	0.133	0.356	0.030	0.241	0.272	0.030	0.012	0.061	0.085	σ_d
		1	0.178	0.374	0.235	0.206	0.084	0.130	0.022	0.035	σ_e
			1	0.077	0.190	0.569	0.034	0.392	0.033	0.096	μ
				1	0.164	0.022	0.235	0.050	0.273	0.113	α
					1	0.361	0.368	0.537	0.087	0.032	n_H
						1	0.357	0.342	0.002	0.061	n_n
							1	0.074	0.144	0.055	hb_{OO}
								1	0.032	0.046	hb_{ON}
									1	0.081	P_G
										1	n_{arf}

For the 3-substituted pyridines (set TM11, Table 20):

$$T_m = -36.7(\pm 10.5)\mu_X + 102(\pm 9.38)n_{hb} + 218(\pm 64.9)\sigma_{lX} - 1500(\pm 211)\sigma_{eX} + 203(\pm 16.0) \quad (107)$$

$100R^2$, 88.87; A$100R^2$, 87.19; F, 37.91; S_{est}, 32.5; S^o, 0.375; N_{dp}, 24.

For the 4-substituted pyridines (set TM12, Table 20):

$$T_m = 241(\pm 90.0)\alpha_X + 119(\pm 11.9)n_{hb} + 123(\pm 48.2)\sigma_{lX} - 584(\pm 186)\sigma_{eX} + 196(\pm 22.0) \quad (108)$$

$100R^2$, 85.47; A$100R^2$, 83.29; F, 27.94; S_{est}, 42.8; S^o, 0.428; n, 24.

The models for the 3- and 4-substituted pyridines are poor when compared with the results for the other data sets studied.

Table 20. T_m of substituted pyridines

Set TM10. 2-Xpy. X, T_m: H, 232; Cl, 227; Me, 206; Et, 210; iPr, 132^x; Pr, 275; Bu, 370^x; CH_2Ph, 286; CH_2CH_2Ph, 270; CH_2CO_2H, 371; $CHPh_2$, 336; $CH_2CH_2CO_2H$, 435; E-2-$VnCO_2H$, 510; Bz, 313; CO_2H, 510; CO_2Me, 316; CO_2Et, 282; $CONH_2$, 405; $CONEt_2$, 298; CN, 325; NO_2, 315; SeCl, 392; SeBr, 389.
Set TM11. 3-Xpy. X, T_m: H, 232; Cl, 212; Br, 246; I, 327; Me, 255; Et, 196; iPr, 227; CH_2Ph, 307; CH_2CO_2H, 417; $CHPh_2$, 352; CCl_3, 263; CH_2OH, 267; $CH_2CH_2CO_2H$, 414; E-2-PhVn, 365; E-2-$VnCO_2H$, 476; Ac, 287; Bz, 317; CO_2H, 410; CO_2Me, 316; CO_2Et, 275; $CONH_2$, 405; CN, 301; NO_2, 344; SH, 354.
Set TM12. 4-Xpy. X, T_m: H, 232; Cl, 230; Br, 274; I, 326; Me, 276; Et, 183; iPr, 218; tBu, 233; CH_2tBu, 321; CH_2Ph, 286; CH_2CO_2H, 404; CH_2OH, 267; $CH_2CH_2CO_2H$, 414; $CHPh_2$, 397; Ph, 351; E-2-$VnCO_2H$, 774; Bz, 344; CO_2H, 592; CO_2Me, 289; CO_2Et, 296; CN, 353; NMe_2, 387; NO_2, 344; SMe, 320.

Note: T_m were taken from refs. 32–35.

Table 21. Correlation matrix for Eq. 106

μ	n_{hb}	υ_2	σ_l	σ_e	
1	0.280	0.061	0.714	0.184	μ
	1	0.082	0.079	0.051	n_{hb}
		1	0.013	0.517	υ_2
			1	0.317	σ_l
				1	σ_e

Table 22. Correlation matrix for Eq. 107

μ	n_{hb}	σ_l	σ_e	
1	0.191	0.793	0.221	μ
	1	0.105	0.139	n_{hb}
		1	0.176	σ_l
			1	σ_e

Table 23. Correlation matrix for Eq. 108

α	n_{hb}	σ_1	σ_e	
1	0.164	0.291	0.305	α
	1	0.084	0.146	n_{hb}
		1	0.120	σ_1
			1	σ_e

C. Critical Temperatures

1. *Introduction*

The temperature at which the phase boundary between liquid and gas disappears is the critical temperature, T_C. We consider the disappearance of a phase to be a phase change property. From the Van der Waals equation of state for gases the relationship:

$$T_C = \frac{8a}{27Rb} \tag{109}$$

for T_C is obtained. R is the gas constant, a is a measure of the intermolecular forces, and b is a measure of the molar volume of the gas. It has been shown that the ratio a/b is a function of a form of the IMF equation.[30] All of the data sets involving critical temperatures, T_C, are reported in Table 20.

2. *Monatomic Species*

The monatomic rare gases (set TC1) were again, correlated with Eq. 21. The regression equation obtained is:

$$T_{C.M} = 2186(\pm 138)\alpha + 66.4(\pm 10.8) \tag{110}$$

$100r^2$, 98.44; F, 252.5; S_{est}, 19.9; S^o, 0.153; N_{dp}, 6.

3. Diatomic Molecules

Correlation of the T_C for homonuclear diatomic molecules (set TC2a) with Eq. 22 gave the regression equation:

$$T_{C,M_2} = 2852(\pm 231)\alpha + 87.2(\pm 28.0) \tag{111}$$

$100r^2$, 96.83; F, 152.7; S_{est}, 54.8; S^o, 0.211; N_{dp}, 7.

The results are slightly improved by the exclusion of T_C for F_2; the regression equation is:

$$T_{C,M_2} = 2979(\pm 224)\alpha + 64.6(\pm 29.4) \tag{112}$$

$100r^2$, 97.79; F, 177.1; S_{est}, 49.1; S^o, 0.182; N_{dp}, 6.

Inclusion of the heteronuclear diatomic molecules in the data set (set TC2) required correlation with Eq. 25 in order to take the dipole moment into account. The regression equation was:

$$T_{C,M^1M^2} = 2806(\pm 197)\alpha + 90.1(\pm 21.0) \tag{113}$$

$100r^2$, 95.30; F, 202.9; S_{est}, 50.3; S^o, 0.237; N_{dp}, 12; $r_{\alpha\mu}$, 0.020.

There is no significant difference between the coefficients of Eqs. 110, 111, and 112.

Table 24. Critical temperatures T_C for sets TC1 through TC5

Set TC1. Monatomic M. M, T_C: He, 5.19; Ne, 44.44; Ar, 150.71; Kr, 209.4; Xe, 289.74; Rn, 376.99.
Set TC2. Diatomic M^1M^2. M^1M^2, T_C: H_2, 33.3; F_2, 144; Cl_2, 417; Br_2, 584; I_2, 785; N_2, 126; O_2, 155; NO, 180; CO, 133; HCl, 325; HBr, 363; HI, 423.
Set TC3. Tetrahedral molecules MX_4. M, X, T_C: C, H, 191; C, F, 227; C, Cl, 556; C, Me, 438; Si, H, 270; Si, Cl, 506; Si, F, 506; Ge, Cl, 551; Sn, Cl, 592.
Set TC4. MeX. X, T_C: H, 191; F, 318; Cl, 416; Br, 464; I, 528; Me, 305; Et, 370; Pr, 425; iPr, 408; tBu, 440; OMe, 406; OEt, 438; OPh, 641; SMe, 503; SEt, 533; OAc, 507; Vi, 365; C_2Me, 489; Ph, 592; CHO, 461; Ac, 510; CO_2Me, 507; CO_2Et, 523; CN, 548; NMe_2, 433; NO_2, 588; CF_3, 346.
Set TC5. PhX. X, T_C: H, 562; F, 560; Cl, 632; Br, 670; I, 721; Me, 594; Et, 617; Pr, 638; iPr, 636; OMe, 642; OEt, 647; OPh, 767; Vi, 648; Ph, 768, 789; CHO, 625; CN, 699; NMe_2, 688.

Note: T_C were taken from refs. 32–35 and 52.

4. *Tetrahedral Molecules*

Correlation of T_C for tetrahedral XM_4 molecules (set TC3) with Eq. 22 on exclusion of the value for CCl_4 gave the regression equation:

$$T_{C,MX_4} = 1177(\pm 54.8)\alpha + 161(\pm 12.3) \tag{114}$$

$100r^2$, 98.71; F, 460.9; S_{est}, 19.5; S^o, 0.131; N_{dp}, 8.

The magnitude of A is much less for these compounds than it is for the rare gases or diatomic molecules. The value of T_C calculated for CCl_4 from Eq. 114 is almost 100° (about five standard deviations) less than that reported in the literature.

5. *Substituted Methanes*

Correlation of T_C for substituted methanes (set TC4) with Eq. 25 gave on exclusion of the value for MeC_2Me the regression equation:

$$T_{C,CH_3X} = 1368(\pm 99.6)\alpha + 52.6(\pm 5.46)\mu + 230(\pm 16.2) \tag{115}$$

$100R^2$, 90.59; A$100R^2$, 90.20; F, 110.7; S_{est}, 31.1; S^o, 0.326; N_{dp}, 26; $r_{\alpha\mu}$, 0.303; C_α, 83.9; C_μ, 16.1.

6. *Substituted Benzenes*

Correlation of T_C for substituted benzenes (set TC5) with Eq. 25 gave the regression equation:

$$T_{C,PhX} = 784(\pm 83.9)\alpha + 21.9(\pm 5.55)\mu + 547(\pm 13.5) \tag{116}$$

$100R^2$, 86.49; A$100R^2$, 85.59; F, 44.83; S_{est}, 23.7; S^o, 0.405; N_{dp}, 17; $r_{\alpha\mu}$., 0.262; C_α, 87.7; C_μ, 12.3.

Although the fit of the data is much poorer for the substituted benzenes the composition of the intermolecular forces is essentially the same with polarizability clearly the predominant factor in determining T_C.

IV. CONCLUSIONS

The IMF model is generally effective in modeling phase change properties. The greatest difficulties are encountered in modeling melting points when changes in the crystal lattice occur within a data set. In view of the quality of the data which have been obtained in many laboratories under various conditions the fit of the data to the model is very good. This is in agreement with previous applications of the IMF model. There is a problem involved in the use of this model, some parameters such as polarizability and steric effect constants, and bond moment and localized electrical effect constants are strongly collinear. It is often impossible to do more than separate the overall substituent effect into polar and nonpolar components. From the point of view of science, which should be explicative, as opposed to engineering, which need not be, this is still a large improvement over methods which make use of "black box parameters".

REFERENCES

1. Grosshans, J. A. *Darstellung der physikalischen Eigenschaften der chemische verbindungen $C_pH_qO_r$ als Funktion der Atomsumme oder Densitätszahl $p + q + r$*, R. Friedländer u. Sohne, Berlin, **1895**.
2. Smiles, S. *The Relations Between Chemical Constitution and some Physical Properties*, Longmans, Green and Co., London, **1910**.
3. Kauffmann, H. *Beziehungen zwischen physikalischen Eigenschaften und chemischen Konstitution*, Verlag von Ferdinand Enke, Stuttgart, **1920**.
4. Kremann, R.; Pestemer, M. *Zusammenhänge zwischen physikalische Eigenschaften und chemischen Konstitution*, Verlag von Theodor Steinkopf, Dresden, **1937**.
5. Charton, M. In *The Chemistry of Dienes and Polyenes*, Rappoport, Z. Ed., Wiley, Chichester, **1997**, pp. 683–732.
6. Saunders, M. R.; Livingstone D. In *Advances in Quantitative Structure Property Relationships*, Charton, M., Ed., JAI Press, Greenwich, Conn., **1996**, pp. 53–79.
7. Burkert, U.; Allinger, N. L. *Molecular Mechanics*, Amer. Chem. Soc., Washington, D. C., **1982**.
8. Martin, Y. C.; Kim Ki-H. In *Advances in Quantitative Structure Property Relationships*, Charton, M., Ed., JAI Press, Greenwich, Conn., **1996**, pp. 1–52.
9. Charton, M. *Topics in Current Chem.* **1983**, *114*, 57–91.
10. Charton, M. *Stud. Org. Chem.* **1992**, *42*, 629–687.
11. Charton, M. *Advances in Molecular Structure Research* **1999**, *5*, 25–88.
12. Kier, L. B.; Hall, L. H. *Molecular Connectivity in Chemistry and Drug Research*, Academic Press, New York, **1976**.
13. Merrifield, R. E.; Simmons, H. E. *Topological Methods in Chemistry*, Wiley-Interscience, New York, **1989**.
14. Balaban, A. T. Ed., *From Chemical Topology to Three-Dimensional Geometry*, Plenum Press, New York, **1997**.
15. Charton, M. *Prog. Phys. Org. Chem.* **1987**, *16*, 287–315.
16. Charton, M. *Topics in Current Chem.* **1983**, *114*, 107–118.
17. Pauling, L. *The Nature of the Chemical Bond*, 3rd Ed, Cornell University Press, Ithaca, New York, **1960**, pp. 221–267, 417–421, 505–540.
18. Charton, M. *Prog. Phys. Org. Chem.* **1971**, *8*, 235–317.
19. Charton, M. *J. Am. Chem. Soc.* **1969**, *91*, 615–618.
20. Charton, M. *Prog. Phys. Org. Chem.* **1973**, *10*, 81–204.
21. Charton, M.; Sirovich, C. *Abstr. 27th M. A. R. M. Am. Chem. Soc.* **1993**, p. 129.
22. Verloop, A.; Hoogenstraaten, W.; Tipker, J. *Drug Design* **1976**, *7*, 165–207.
23. Charton, M.; Charton, B. I. *J. Theoret. Biol.* **1982**, *99*, 629–644.
24. Charton, M. *Prog. Phys. Org. Chem.* **1990**, *18*, 163–284.

25. Charton, M. In *The Chemistry of Organic Germanium, Tin and Lead Compounds*, Patai, S., Ed., Wiley, Chichester, **1995**, pp. 603–664.
26. Charton, M.; Charton, B. I. *Abstr. Intl. Symp. Lipophilicity in Drug Research and Toxicology*, Lausanne, **1995**, pp. O–3.
27. Charton, M. In *Classical and 3-D QSAR in Agrochemistry and Toxicology*, Hansch C.; Fujita, T.; Eds., American Chemical Society, Washington, D.C., **1995**, pp. 75–95.
28. Charton, M.; Charton, B. I. *J. Phys. Org. Chem.* **1994**, *7*, 196–206.
29. Charton, M. In *Trends in Medicinal Chemistry '88*, van der Goot, H.; Domany, G.; Pallos L.; Timmerman, H.; Eds., Elsevier, Amsterdam, **1989**, pp. 89–108.
30. Charton, M.; Charton, B. I. *J. Org. Chem.* **1979**, *44*, 2284–2288.
31. Miller, T. M. In *Handbook of Chemistry and Physics*, 67th Edn., Weast, R. C.; (Astle M. J., Beyer, W. H.; Eds.), CRC Press, Boca Raton, **1986**, pp. E66–E75.
32. Buckingham, J.; Macdonald, F. Exec. Eds. *Dictionary of Organic Compounds*, 6th Edn., Chapman and Hall, London, **1996**.
33. Weast, R. C.; (Astle, M. J.; Beyer, W. H.; Eds.), *Handbook of Chemistry and Physics*, 67th Edn., CRC Press, Boca Raton, **1986**.
34. Dean, J. A. Ed., *Handbook of Organic Chemistry*, McGraw-Hill, New York, **1987**.
35. Budavari, S. Ed., *The Merck Index*, 11th Edn., Merck and C., Rahway, N. J., **1989**.
36. Charton, M. In *The Chemistry of Organic Germanium, Tin and Lead Compounds, Supplement A*, Rapaport, Z.; Ed., Wiley, Chichester, in press.
37. The regression equation given in ref. 5 is incorrect.
38. Majer, V.; Svoboda, V. *Enthalpies of Vaporization of Organic Compounds. A Critical Review and Data Compilation*, Blackwell Scientific Publicaitons, Oxford, **1985**.
39. Chickos, J. S.; Braton C. M.; Hesse D. G.; Liebman, J. F. *J. Org. Chem.* **1991**, *56*, 927–938.
40. Charton, M. *Prog. Phys. Org. Chem.* **1981**, *13*, 119–251.
41. Robertson, J. M. *Organic Molecules and Crystals*, Cornell University Press, Ithaca, **1953**, pp. 183–218.
42. Hall, G. G. *Molecular Solid State Physics*, Springer Verlag Berlin, Heidelberg, **1991**, pp. 41–45.
43. Nishio, M.; Hirota, M.; Umezawa, Y. *The CH/π Interaction. Evidence, Nature, and Consequences*, Wiley-VCH, New York, **1998**, pp. 53, 134, 196–198.
44. Umezawa, Y.; Tsuboyama, S.; Honda, K.; Uzawa, J.; Nishio, M. *Bull. Chem. Soc. Jpn.* **1998**, *71*, 1207–1213.
45. Clar, E. *Polycyclic Hydrocarbons*, Vol. 1, Academic Press, London, **1964**, pp. 126–129.
46. McClellan, A. L. *Tables of Experimental Dipole Moments*, W. H. Freeman, San Francisco, **1963**.
47. McClellan, A. L. *Tables of Experimental Dipole Moments Vol. 2*, Rahara Enterprises, El Cerrito, Cal **1974**.
48. Charton, M. *Advances in Quantitative Structure Property Relationships* **1996**, *1*, 171–219.
49. Charton M. In *The Chemistry of Sulfenic Acids, Esters and Derivatives*, Patai, S., Ed., Wiley, New York, **1990**, pp. 657–700.
50. Charton M. In *The Chemistry of Arsenic, Antimony and Bismuth*, Patai, S., Ed., Wiley, New York, **1994**, pp. 367–439.
51. Clar, E. *Polycyclic Hydrocarbons*, Vol. 1 and 2, Academic Press, London, **1964**.
52. Reid, R. C.; Prausnitz J. M.; Poling, B. E. *The Properties of Gases and Liquids*, McGraw-Hill, New York, **1987**.

APPENDIX I

Abbreviations:

Me	methyl	Ted	tetradecyl
Et	ethyl	Pdc	pentadecyl
Pr	1-propyl	Hxd	hexadecyl
iPr	2-propyl	Hpd	heptadecyl
Bu	butyl	Ocd	octadecyl
iBu	2-methyl-1-propyl	Ndc	nonadecyl
tBu	2-methyl-2-propyl	Ec	eicosanyl
Pe	pentyl	Ph	phenyl
Hx	hexyl	Pn	phenylene
cHx	cyclohexyl	Nh	Naphthyl
Hp	heptyl	Nn	Naphthalene
Oc	octyl	Ant	anthracenyl
Nn	nonyl	Pht	phenanthracenyl
Dc	decyl	Py	Pyridyl
Udc	undecyl	Vi	vinyl
Ddc	dodecyl	1-Vn	vinylidene
Trd	tridecyl	2-Vn	vinylene
Ac	acetyl	Ak	alkyl
Bz	benzoyl	C_2H	ethynyl
Ts	4-toluenesulfonyl	C_2	ethynylene
Unc	uncosanyl	Doc	docosanyl
Trc	tricosanyl	Tec	tetracosanyl
Pec	Pentacosanyl	Hxc	Hexacosanyl
Hpc	Heptacosanyl	Occ	octacosanyl
Noc	nonacosanyl	Tc	triacontanyl

INDEX